SpringerBriefs in Food, Health, and Nutrition

For further volumes:
http://www.springer.com/series/10203

Valérie Guillard · Claire Bourlieu
Nathalie Gontard

Food Structure and Moisture Transfer

A Modeling Approach

 Springer

Valérie Guillard
Nathalie Gontard
UMR 1208 Agropolymers Engineering
  and Emerging Technologies
Montpellier cedex 5
France

Claire Bourlieu
INRA Agrocampus
UMR 1253 Science and Technology
  of Milk
Rennes
France

ISBN 978-1-4614-6341-2   ISBN 978-1-4614-6342-9   (eBook)
DOI 10.1007/978-1-4614-6342-9
Springer New York Heidelberg Dordrecht London

Library of Congress Control Number: 2013930688

Printed on acid-free paper

Springer is part of Springer Science+Business Media (www.springer.com)

Contents

Abstract

The physical state and structural characteristics of food materials influence their mass transfer properties, especially during most processing units, such as drying, hydration, storage, and so forth. All sorts of structural characteristics coexist according to the type of food material. In porous cereal-based products, the effective water diffusivity is highly affected by the volume fraction and the distribution of both the solid and gas phases, while in homogeneously dense food materials, such as fat-based coatings or other edible coatings, the effective water diffusivity depends mainly on factors that affect the "tightness" of the molecular structure (i.e., free volume, cohesive energy density, crystallinity, etc.).

In this book, the impact of the structure of food, including edible coatings, on mass transfer properties is reviewed and illustrated in the challenging case of moisture transfer, for which studies are widely available. The first part is devoted to the theoretical background necessary to understand and achieve a complete characterization and modeling of moisture transfer in food. Then, in the second part, a multi-scale analysis of the structure/moisture-transfer relationship is proposed, focusing first on the molecular structure (e.g., free volume, crystallinity), then on the nanoscale structure, and finally on the microscale (e.g., porous food) and macroscale (e.g., bilayer) structures. The continuity of knowledge between these different scales will be analyzed and illustrated in the case of fat-based edible coatings. For each scale of structural observation, a focus on the mathematical modeling of the relationship between structural properties and moisture transfer properties will be performed. In regard to moisture transfer properties, equilibrium (water sorption isotherm) and dynamic (diffusivity) water transfer properties will be considered.

Introduction

In food science and engineering, the control of moisture content and transfer is one of the main challenges in anticipating and preventing food decay and subsequent food losses during processing and storage. Moisture is indeed involved in all the main reactions of food degradation (e.g., microorganisms' growth, enzymatic hydrolysis, texture losses, etc.). Lowering the moisture content of food (and more precisely, its water activity) permit the reduction of occurrence of an unwanted reaction of food degradation. Preventing food moisture gain or loss from the surrounding atmosphere avoids the softening of a crispy dry texture or, on the contrary, the hardening of a soft moist texture, respectively. That's why moisture transfer has been intensively studied and modeled in the last century, especially drying, in order to properly dimension driers, to preserve food quality during drying, to determine optimal final water activity of the dry product to ensure its long-term stability, and to optimize energy consumption during drying. Indeed, drying is the more energy-consuming processing unit used in the food industry.

Moisture transport in solid food products can be considered as a diffusion-controlled process. Diffusion is described by Fick's law, in which the mass flux is linear with the gradient in moisture content according to a coefficient of diffusion, also called *diffusivity*. This diffusivity characterizes the ability of water to diffuse through the food matrix. The moisture transport is strongly influenced by the structure of the food product, mainly through this value of diffusivity. This value may change according to the texture (e.g., porous vs. dense matter), according to the formulation (e.g., presence of high content of fat or salt), and according to the moisture content itself. For example, in porous cereal-based products, the water diffusivity is highly affected by the volume fraction and the distribution of both solid and gas phases, while in dense food materials, such as fat-based coatings or other edible coatings, it depends on factors related to the molecular structure (i.e., free volume, cohesive energy density, crystallinity, etc.).

Although the relationship between values of water diffusivity and structural parameters of matter was early evidenced, the formalization (i.e., modeling) of this relationship has scarcely been attempted in the food science sector. More often, than not, this formalization has been done using empirical equations that can rarely be extrapolated to a system different from the one used to establish them.

V. Guillard et al., *Food Structure and Moisture Transfer*, SpringerBriefs in Food, Health, and Nutrition, DOI: 10.1007/978-1-4614-6342-9_1, © The Author(s) 2013

On the contrary, in polymer science, tortuosity-based mathematical models have been used for decades to take into account the impact of the addition of water-resistant fillers in the matrix on water diffusivity. Contrary to polymers, food is subjected to the natural complex behavior of the biological matter from which it derives. This complexity may explain why such a modeling approach of the structure/mass-transfer relationship has scarcely been attempted.

The aim of this brief review is to analyze the main results of the scientific literature, pointing out the relationship between food structure and moisture transport properties. The first section recalls the theoretical background concerning moisture transport properties (solubility and diffusivity) and their determination. The main theories and approaches used for modeling moisture transport will be presented, and the predominance of Fick's law among all of them will be justified. In a second part, a multiscale analysis of the structure/mass-transport relationship is proposed, focusing first on the molecular structure (e.g., free volume, crystallinity), then on the nanoscale structure, and finally on the microscale (e.g., porous food) and macroscale (e.g., bilayer) structures. The continuity of knowledge between these different scales will be analyzed and illustrated in the case of fat-based edible coatings. For each scale of structural observation, a focus on the mathematical modeling of the relationship between structural properties and moisture transfer properties will be performed. Then, in a third part, an example of an integrated multiscale analysis will be presented and illustrated in the case of edible lipid-based films.

Theoretical Background

Water Vapor Sorption Isotherms

The term *sorption* is generally used to describe the initial penetration and the dispersal of the permeant molecules into food matrices (Chirife and Iglesias 1978) and, more generally, into all kinds of polymers (Naylor 1989; Tsujita 1992). This term includes adsorption, absorption into microvoids, and cluster formation. When a food matrix is placed in a humid atmosphere, the relationship between the ambient water activity (a_w) and the water concentration (Q_w) in the material at a given temperature is described by an equilibrium sorption isotherm $(Q_w = f(a_w))$. Water concentration (Q_w) can be defined as the ratio of the mass of sorbed water at equilibrium (M_w) to the mass of dry matter (M_d):

$$Q_w = \frac{M_w}{M_d} \tag{1}$$

The water activity a_w at a given temperature can be defined as the ratio of the water vapor pressure (p_w) to the saturated water vapor pressure $(p_{w,0})$:

$$a_w = \frac{p_w}{p_{w,0}} \tag{2}$$

Water sorption models are mathematical equations (linking Q_w and a_w or Q_w and p_w) used for the prediction of the sorption properties of materials and for analyzing sorption mechanisms and possible interactions between the substrate and water. In some cases, the parameters involved in these equations have physical meaning and can provide useful information on the possible interactions between the product and water (e.g., the dimensionless solvent-polymer interaction parameter χ in the Flory–Huggins theory) or on the physical state of the substrate (e.g., amorphous, crystalline, etc.).

V. Guillard et al., *Food Structure and Moisture Transfer*, SpringerBriefs in Food, Health, and Nutrition, DOI: 10.1007/978-1-4614-6342-9_2, © The Author(s) 2013

Measurement of Water Sorption Isotherms

The measurement of water sorption isotherms requires bringing the material to an equilibrium state corresponding to a point on the sorption curve and measuring its moisture content when a_w is fixed or, conversely, measuring an a_w when moisture content is controlled (Bell and Labuza 2000). In both cases, the method involves changing the total amount of water in the product being studied by imposing a moisture transfer between the air surrounding the product and the product itself. Once equilibrium is reached, the a_w or the moisture content can be measured using manometric, hygrometric, or gravimetric methods. Manometric methods measure the vapor pressure of water (p_w) in equilibrium with the food material by using sensitive manometers. Hygrometric methods measure the equilibrium *relative humidity* (RH) of air in contact with the food material by using dew point or electric hygrometers. Gravimetric methods involve the registration of sample weight changes. The most common technique, recommended by the European COST action 90 on physical properties of foodstuffs, uses thermostated jars filled with saturated salt solutions to set the air RH. The evolution of weight of the sample with time is recorded until the equilibrium is reached (Lomauro et al. 1985a, b).

Hydrous equilibrium is reached very slowly when water activity is high (Timmermann and Chirife 1991; Vos and Labuza 1974). This slowness favors microbiological and biochemical changes in the product that can lead to the generation of small solutes with high water-binding capacity and disturb hydrous equilibrium. To avoid microbial growth, sodium azide, phenyl mercury acetate, or thymol can be added at low concentrations (about 0.05 % of weight), but the consequences on the product and, thus, water binding, are difficult to assess. In addition to these problems of equilibrium time, the detection of the hydrous equilibrium point can be difficult, since many products exhibit two states of equilibrium for the same water content (the so-called hysteresis phenomenon; Simatos 2002). To avoid these two major problems and, particularly, too long an equilibrium time, alternative methods have been developed.

In the last decade, the water sorption of food products has been evaluated using controlled atmosphere microbalance and dynamic automated sorption methods (Fig. 1). The small size of samples and the dynamic airflow around the samples enables the generation of a complete isotherm (from 0 to 95 % RH) in less than a week. Several works using controlled atmosphere microbalance have been published for sorption isotherm measurement in cereal-based products, such as sponge cake (Guillard et al. 2003a; Roca et al. 2008), dry biscuit (Guillard et al. 2004b), or wafer (Bourlieu et al. 2006), and dense materials, such as lipid-based edible films (Bourlieu et al. 2006, 2008, 2009a, 2010) or wheat gluten materials (Angellier-Coussy et al. 2011). Another advantage of this technique is that the mass change kinetics is recorded as a function of time for each water activity tested and permits identifying some water vapor diffusivity in the material (see the section below on water diffusivity and the work of Guillard et al. 2003c). A drawback of the technique is in determining precisely the equilibrium at a given a_w value for highly

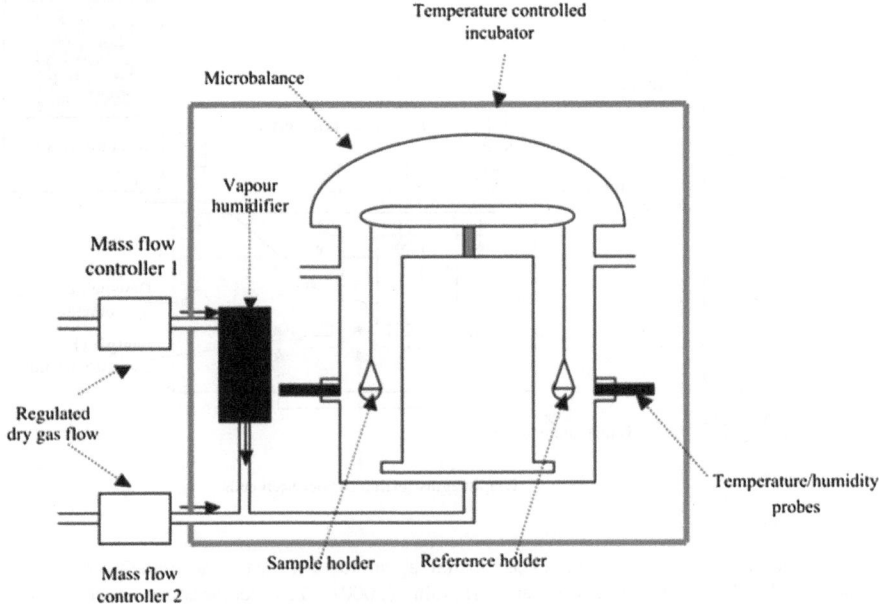

Fig. 1 Example of sorption balance inspired from the Dynamic Vapor Sorption system of Surface Measurement Systems (Alperton, Middlesex, UK)

hydrophobic material that sorbs very low amount of water (lower than the sensitivity of the balance). In this case, the length of the step of equilibration at a given humidity should not be automatically finished when the variation in sample mass is lower than a given value (generally ±0.001–0.002 % total weight/min). This mass criterion is replaced by a forced length of equilibration, which can vary between 8 h below 60 % RH and reach 24 h above this value for very hydrophobic materials (Bourlieu et al. 2006).

Whatever the matrix, in order to gain more precision of the water sorption at very high water activity (>0.95), which is difficult in practice, Baucour and Daudin (2000) have developed a new method whose principle is described in Fig. 2. The apparatus is composed of a set of 10 thermostated cells in which the samples are placed. By using initially saturated air at a given temperature in the first cell and by making successive water vapor pressure drops from cell to cell, a specific range of air RH can be covered (ranging from 98 to 88 % RH over the 10 cells). The principle of this technique is based on the fact that an isothermal pressure drop of 1,000 Pa in an initially saturated air at atmospheric pressure lowers its RH to 99 %, while a variation of 10^4 Pa will cause only a negligible change in a_w of the product, showing that modulation of atmospheric pressure is a much more precise way of adjusting a set of high RH levels. Contrary to saturated salt solutions, this pressure regulation allows the realization of a sequence of RH drops of very low amplitude (1 %) and of high accuracy. Moreover, thin slices of material are submitted to an air flow of very high velocity (>10 m/s), which allows a significantly reduction in the

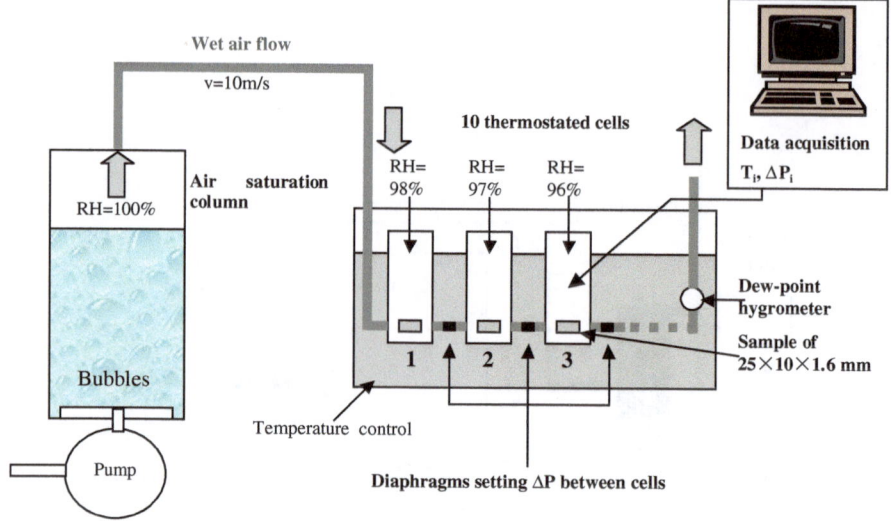

Fig. 2 Schematic diagram of the experimental apparatus for fast measurement of sorption isotherms developed by Baucour and Daudin (2000). Ti = temperature in each cell, ΔPi = pressure difference in each cell, v = wet air flow velocity. Adapted from Baucour and Daudin (2000)

equilibrium time. Baucour and Daudin successfully validated their method with gelatine gel samples, by comparison with the standard saturated salt method. However, this method was not applied to very hydrophobic materials.

Mathematical Description of Water Sorption Isotherm

Water vapor sorption isotherm in food is first obtained under the form of sets of experimental data. Many different models have been reported for the analysis of sorption of water in various food and nonfood materials (Park 1986; Khalfaoui et al. 2003). For example, in the case of nonedible matrices such as polymers, the classification of Rogers (1965) highlights five fundamental sorption modes: Henry, Langmuir, dual mode, Flory–Huggins, and type IV; the latter has a sigmoid shape and reflects a complex combination of several sorption modes (Fig. 3) (Rogers 1965; Tien 1994). Most food matrices present the Flory–Huggins or type IV shape of isotherms. The type IV is specifically representative of matrices rich in hydrophilic macromolecules, such as cereal-based products. This specific classification, developed for water sorption in polymers, joins that of Brunauer et al. (1940), who first proposed a systematic attempt to interpret adsorption isotherms for any kinds of gas–solid equilibria. Brunauer et al. (1940) classified isotherms into five types (from type I to type V). Type I isotherms characterize microporous

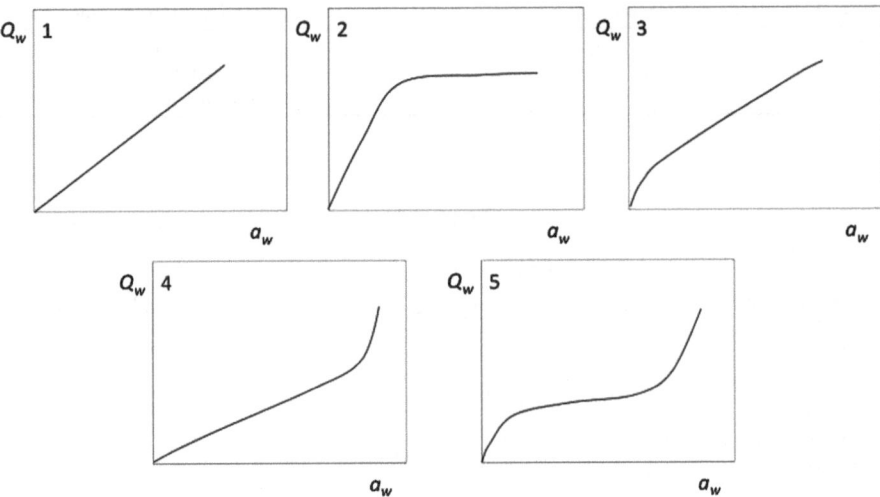

Fig. 3 Schematic presentation of the types of gas and vapor sorption in the polymer according to the classification of Rogers (1965): (1) Henry's law; (2) Langmuir; (3) Dual-mode; (4) Flory–Huggins; (5) type IV

adsorbents (Langmuir mode). Types II and III describe adsorption on macroporous adsorbents with string and weak adsorbate–adsorbent interactions, respectively. Types IV and V represent adsorption isotherms with hysteresis.

Water sorption isotherm data are almost always fitted with one or more of a host of theoretical and empirical models (Table 1). No single equation has been found to depict accurately the sorption isotherms of all types of foods in the entire range of water activity. As a consequence, isotherm data are studied individually, and the model that fits and describes most accurately the experimental data is used. Table 1 reports the main equations currently encountered in the modeling of moisture sorption isotherms in food science and their range of validity.

Several common mathematical models for predicting water sorption in food and in nonfood materials are based on the multilayer sorption theory of small molecules onto solid surfaces. These multilayer sorption models combine two modes of sorption. The first sorption mode comprises the formation of a monolayer of sorbate molecules on the surface of the sorbent. The second sorption mode is the multilayer condensation of the sorbate onto the sorbate monolayer, whereby the properties approach those of the pure liquid. The best known example of such multilayer sorption isotherm models is the BET (Brunauer et al. 1938; De Boer 1953) equation in which the multilayer condensation is combined with the Langmuir sorption model for the formation of the monolayer (Table 1).

Whereas the physical interpretation of surface sorption onto rigid adsorbents following the BET equation is rather clear (Adamson 1967), the interpretation of the sorption of water by biopolymer matrices is less evident, since for these systems, the surface adsorption of water is much smaller than the absorption of

Table 1 Examples of equations used in the modeling of food water sorption isotherms

Reference	Equation[a]	Unknown adjustable parameters	a_w range of validity
Brunauer et al. (1938)	$Q_w = \frac{Q_{mono}Ca_w}{(1-a_w)(1-a_w+Ca_w)}$	Q_{mono}, C	0.3–0.4
Oswin (1946)	$Q_w = c_1\left(\frac{a_w}{1-a_w}\right)^{c_2}$	c_1, c_2	0.3–0.5
Smith (1947)	$Q_w = c_1 - c_2\ln(1-a_w)$	c_1, c_2	0.3–0.5
Halsey (1948)	$Q_w = \left(\frac{c_2}{\ln a_w}\right)^{-1/c_1}$	c_1, c_2	0.1–0.8
Anderson (1946) De Boer (1953) Guggenheim (1966)	$Q_w = \frac{Q_{mono}CKa_w}{(1-Ka_w)(1-Ka_w+CKa_w)}$	Q_{mono}, C, K	<0.94
Chirife et al. (1983) Ferro Fontan et al. (1982)	$Q_w = \left[\ln\left(\frac{a}{a_w}\right)\frac{1}{b}\right]^c$	a, b, c	>0.85
Timmerman and Chirife (1991)	$Q_w = \frac{CKQ_{mono}a_wH'(h)H(h)}{(1-Ka_w)(1-Ka_w+CH(h)Ka_w)}$	Q_{mono}, C, K, h	0–1
Peleg (1993)	$Q_w = k_1a_w^{n_1} + k_2a_w^{n_2}$	k_1, k_2, n_1, n_2	
Viollaz and Rovedo (1999)	$Q_w = \frac{CkQ_{mono}a_w}{(1-ka_w)(1-ka_w+Cka_w)} + \frac{Ckk_2a_w^2}{(1-ka_w)(1-a_w)}$	Q_{mono}, C, K, k, k_2	0–1

[a] Nomenclature
Q_w moisture content in g/g (dry basis)
Q_{mono} moisture content at the monolayer value g/g (dry basis)
C energy constant
K additional parameter to the BET equation
H and H' additional functions containing a 4th parameter (h) to the GAB equation
k_1, k_2 constants in Peleg equation
a, b, c constants
$n_1 > 1$, $n_2 > 1$

water by the bulk of the material, even at low water contents. In addition, the BET equation fails to take into account the glass transition of the matrix and the molecular interactions of macromolecules and water in the rubbery state.

Experience has shown that the BET equation usually fits poorly the water sorption by biopolymer matrices at high water activities ($a_w > 0.4$). For this reason, the so-called Guggenheim–Anderson–de Boer (GAB) isotherm equation was introduced and is often applied to fit and analyze water vapor sorption by biopolymer matrices (Anderson 1946; Roos 1995). The GAB equation is obtained from the BET equation by transformation of the water activity axis:

$$a_w \rightarrow Ka'_w \tag{3}$$

Then,

$$Q_w = \frac{Q_{mono}CKa'_w}{\left(1 - Ka'_w\left(1 - Ka'_w + CKa'_w\right)\right)} \tag{4}$$

It should be noted that the physical significance of the GAB equation for the sorption of water by biopolymer matrices is even more limited than that of the BET equation, because of its unphysical divergence at water activity values below unity for values of K larger than 1. Indeed, whereas the BET equation diverges as $(1 - a_w)^{-1}$ for $a_w \rightarrow 1$, the GAB equation diverges at a critical water activity $a_w = K^{-1}$. This unphysical divergence was observed, for example, for the GAB fit of all maltopolymer-maltose samples studied in the work of Ubbink et al. (2007).

Q_{mono} is usually interpreted as the water content at which a full monolayer of sorbate coverage is reached, and in the relevant limit of $C \gg 1$, C determines the steepness of the initial rise to the monolayer plateau and is thus related to the free energy of sorption at low partial pressures. It was hypothesized that both parameters are thus related to the glassy-state properties of the matrices (Ubbink et al. 2007). In contrast, as was suggested by Ubbink et al (2007), K should be related either to the rubbery-state properties or to the glass transition of the matrices, since K^{-1} equals the water activity at which the GAB equation diverges.

The GAB equation was found to represent adequately the experimental sorption data in many foodstuffs, such as amorphous maltodextrin–glycerol matrices (Roussenova et al. 2010), carbohydrate mixtures (Ubbink et al. 2007), dried fruits (Maroulis et al. 1988), tapioca (Sanni et al. 1997), cookies and corn snacks (Palou et al. 1997), and hydrophobic films (Bourlieu et al. 2006, 2009). However, in the very high a_w range ($a_w > 0.94$), which is of most interest for fresh foods, the GAB equation failed to describe the water vapor sorption. Alternative empirical models, such as the Peleg (1993) or the Ferro-Fontan equations (Ferro-Fontan et al. 1982; Chirife et al. 1983) were consequently proposed and were found adequate to predict moisture content in the very high a_w range in products such as gelatine gel (Baucour and Daudin 2000) or sponge cake (Guillard et al. 2003a).

Water Sorption Behavior in Foodstuffs

The equilibrium moisture sorption isotherms of hydrophilic food products usually show a large degree of upturn at high activities. Two phenomena are suggested for interpretation of the equilibrium isotherm upturn in dense materials: (1) plasticization of the material by water (when the product, initially in a glassy state, passes through its glass transition and reaches the rubbery state) and/or (2) clustering of water molecules. An in-depth analysis of water sorption isotherms allows analyzing this phenomenon of water clustering, as described below.

Water molecules have a tendency to form clusters when absorbed in a polymer. To provide a measure for clustering, Zimm and Lundberg (1956) defined the clustering integral $G_{w,w}$. The ratio of the clustering integral to the partial molecular volume, $G_{w,w}/V_w$, called the *clustering function*, is obtained from the equilibrium data:

$$\frac{G_{w,w}}{v_w} = -(1 - \phi_w) \left[\frac{\partial(a_w/\phi_w)}{\partial a_w} \right]_{P,T} - 1 \qquad (5)$$

where v_w is the partial molecular volume for water. Positive values of the clustering function indicate that water in the polymer matrix forms small clusters, or "pockets," between the polymer chains. Conversely, negative values of the clustering function indicate that water is molecularly dispersed in the polymer matrix. The state $G_{w,w}/v_w = -1$ represents a random distribution of noninteracting water molecules. The quantity $\phi_w G_{w,w}/v_w$ has been interpreted as the mean number of solvent molecules in the neighborhood of a given solvent molecule in excess of those provided by the mean solvent concentration. The sum $\left(1 + \phi_w G_{w,w}/v_w\right)$, therefore, is the mean cluster size.

Water Diffusion

Moisture transfer in foodstuffs and, more particularly, in multidomain systems occurs because of a difference in water activity between domains, which acts as a driven force for the mass transfer. Moisture gain or loss from one region to another region will continuously occur in order to reach thermodynamic equilibrium between the different domains involved. The movement of water within solids is complex and may be explained by different mechanisms, such as molecular diffusion of liquid water due to concentration gradient, liquid movement due to capillary flow, liquid movement due to gravity, vapor diffusion due to partial vapor pressure, Knudsen diffusion, and surface diffusion (Barbosa-Canovas and Vega-Mercado 1996) (see Fig. 4). In order to explain, describe, and predict moisture movements within solid foods, mathematical models of increasing complexity have been proposed in the literature for decades. In the following section, we will provide an overview of these modeling approaches, discuss their pros and cons, and make conclusions about the most useful ones.

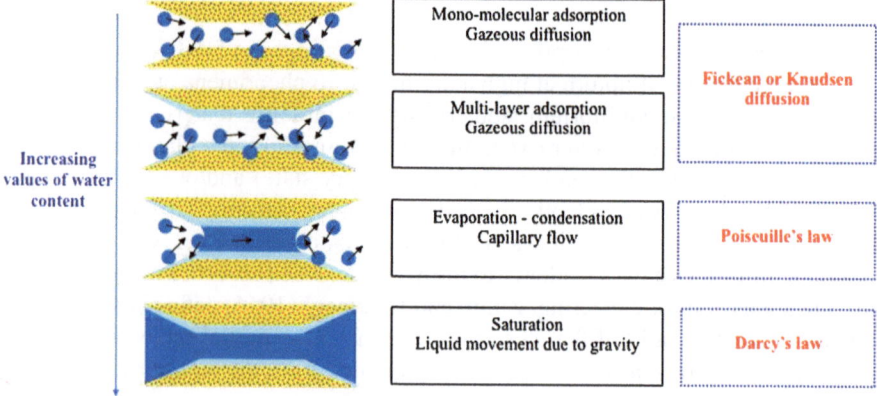

Fig. 4 Schematic presentation of the mechanisms of water transport in a foodstuff contributing to the overall water movement

Modeling Moisture Transfer

Modeling moisture transfer within solid food products has been widely studied because it is necessary for the dimensioning of food-processing units such as drying or storage. Models encountered in the literature can be classified as theoretical models, semitheoretical models, phenomenological models, the Stefan–Maxwell approach, and empirical models according to the assumptions made for internal water transport mechanisms. In Table 2, some of the main mathematical theories are presented according to this classification. For a more exhaustive list of mathematical models, see the review of Waananen et al. (1993), who reported the drying theories encountered in the literature from 1907 to 1992.

1. Theoretical or Mechanistic Models

Theoretical models (or mechanistic models) presuppose the knowledge of the mechanisms of water transport involved in the product being studied—that is, the fluxes and the driving forces involved. In order to tentatively explain the contribution of several mechanisms to an overall water movement, a combination of several driving forces could be used. For example, the "Philip and De Vries" or "Berger and Pei" theory combines liquid capillarity and vapor diffusion (Table 2). Thorvaldsson and Janestad (1999) tentatively tried to take into account liquid and vapor water diffusion in their model and developed a model based on Fick's second law, including two diffusion coefficients for liquid and vapor water, respectively. They successfully validated their model with bread slabs during drying. But in a general manner, such combinations complicate the partial differential equation of diffusion a lot. Difficulty was also encountered in determining separately liquid and water vapor diffusion coefficients. As a result, a simple diffusion model based on Fick's second law (Eq. 6) is usually preferred by investigators to represent the overall moisture transfer in solid foodstuffs, even though, in this case, Fick's law is extrapolated out of its application conditions:

$$\frac{\partial Q}{\partial t} = \frac{\partial}{\partial x} D \frac{\partial Q}{\partial x} \qquad (6)$$

where D is the coefficient of diffusion or diffusivity ($m^2 \ s^{-1}$), Q is the concentration (kg m^{-3}), x is the distance (m), and t is time (s).

These theoretical models based on Fick's second law yield to a good description and prediction of the moisture transfer between the product and the surrounding medium, provided that the assumptions and parameters required in the model formulation were verified and known (sample geometry, moisture diffusivity value, initial and boundary conditions, etc.). Fick's second law has been extensively and successfully used for describing and predicting various processes, such as, for example, soaking of soybeans (Hsu 1983), drying of corn extruded pasta (Andrieu and Stamatopoulos 1986), moisture distribution in dough/raisin mixtures (Karathanos and Kostaropoulos 1995), dehydration of prune (Sabarez et al. 1997; Sabarez and Price 1999), or drying of cashew kernel (Hebbar and

Table 2 Some of the main theories proposed for water transport modeling in food products

Model	Reference	Driving force	Postulated mechanisms	Mass transport equation
Theoretical models	Fick (1855)	∇X_L	Liquid diffusion	$\frac{\partial X}{\partial t} = \frac{\partial}{\partial x}\left(D_{\text{eff}} \frac{\partial X}{\partial x}\right)$
	Okozuno and Doi (2008)	∇X_L	Liquid diffusion + stress-driven diffusion term	
	Colon and Aviles (1993)	∇P	Liquid capillarity	$\frac{1}{A}\frac{\partial X}{\partial t} = -K_H \nabla P$
	Philip and De Vries (1957)	$\nabla X_V, \nabla T, \nabla P$	Liquid capillarity + vapour diffusion	$\frac{\partial X}{\partial t} = \nabla \cdot (D_m \nabla X) + \nabla \cdot (K_{Tm} \nabla T) + \frac{\partial K_H}{\partial z}$
	Berger and Pei (1973)	$\nabla X_L, \nabla X_V$	Liquid capillarity + liquid diffusion + vapour diffusion	$D_L \rho_L \frac{\partial^2 X}{\partial x^2} + D_V\left[(\varepsilon - X)\left(\frac{\partial^2 \rho_w}{\partial x^2}\right) - \left(\frac{\partial X}{\partial x}\right)\left(\frac{\partial \rho_w}{\partial x}\right)\right]$ $= (\rho_L - \rho_w)\left(\frac{\partial X}{\partial t}\right) + (\varepsilon - X)\left(\frac{\partial \rho_w}{\partial t}\right)$
	Whitaker (1980)	$\nabla X_L, \vec{g}, \nabla X_V, \nabla P$	Bulk flow	$\frac{\partial\left(\Psi_\gamma \langle\rho\rangle_\gamma\right)}{\partial t} + \nabla \cdot \left(\langle\rho_1\rangle_\gamma \langle v\rangle_\gamma\right) + \frac{1}{v}\int \rho_l(v_l - w)\cdot\overrightarrow{n}_{\gamma l} dA$ $= \nabla \cdot \left(\langle\rho\rangle_\gamma \cdot D_{\text{eff}\gamma} \cdot \nabla\langle\rho_1\rangle_\gamma / \langle\rho_\gamma\rangle_\gamma\right)$
	Kerkhof (1994)	∇X_{avg}	Bulk flow	$F_s X_{\text{in}} - F_{\text{sout}} X_{\text{out}} = \frac{d(M_p X_{\text{avg}})}{dt}$
	Thorvaldsson and Janestad (1999)	$\nabla X_V, \nabla X_L$	Liquid diffusion + vapour diffusion	$\frac{\partial X_V}{\partial t} = \frac{\partial}{\partial x}\left(D_V \frac{\partial X_V}{\partial x}\right)$ and $\frac{\partial X_L}{\partial t} = \frac{\partial}{\partial x}\left(D_L \frac{\partial X_L}{\partial x}\right)$
Semi-theoretical	Henderson and Pabis (1961)	[–]	Bulk flow	$MR = a\exp(-kt)$
Phenomenological	Henderson (1974)	[–]	Bulk flow	$MR = a\exp(-k_1 t) + b\exp(-k_2 t)$
	Bruce (1985)	[–]	Bulk flow	$MR = \exp(-kt)$
	Luikov (1966, 1975)	$\nabla X, \nabla T, \nabla P$	Vapour diffusion, bulk flow, liquid	$\vec{J}_i \sum L_{ik} \bar{X}_i$
Stefan-Maxwell	Gekas (1992)	∇X_i	Bulk flow	$\frac{\Delta x_i}{x_i} = \sum_j x_j \frac{\bar{u}_i - \bar{u}_i}{k_{ij}}$
Empirical models	Thompson et al. (1968)	[–]	[–]	$t = a\ln MR + b(\ln MR)^2$
	Wang and Singh (1978)	[–]	[–]	$MR = 1 + at + bt^2$

(continued)

Table 2 (continued)

Nomenclature

$\langle\rangle$	Average value	$\vec{J}_i$	Heat and mass diffusion flux (Luikov's theory)
α	The volume fraction of air in the pores	$\overline{X}_i$	Thermodynamic forces giving rise to $\vec{J}_i$ (Luikov's theory)
ε	Void fraction of the solid	A	Exposed area
ρ	Density	D_{effv}	Gas phase effective diffusivity
τ	Factor taking into account the tortuosity	D_{eff}	Effective diffusivity (Fick's law)
v	Individual mass velocity (Whitaker theory)	D_L	Liquid conductivity
Ψ	Volume fraction of the phase	D_m	Overall isothermal moisture diffusivity
∇	Gradient	D_V	Vapour diffusion coefficient
a	Constant	F_s	Solids feed
b	Constant	G	Dry air stream
k_1	Constant	K_H	Unsaturated hydraulic conductivity
k_2	Constant	K_{Tm}	Overall thermal moisture diffusivity
k_{ij}	Mass transfer coefficient	L_{ik}	Luikov phenomenological coefficient
1	Evaporating species (Whitaker theory)	M_p	Dry solid holdup in the dryer
$\vec{n}$	Unit normal vector	MR	Moisture ratio $(X - X_e/X_0 - X_e)$ with X_0, initial moisture content and X_e, equilibrium moisture content
		P	Pressure
T	Time	T	Temperature
u	Velocity	V	Dimensionless heat parameter
x	Spatial dimension	X	Moisture content in kg of water/kg of dry solids
$\overline{x}_i$	Arithmetic mean mole fraction of a component i	Y	Air moisture content
		Z	Coordinate where mass transfer occurs

Subscripts

S	Solid phase
fines	Particules carried by the air
γ	Gas phase
W	Vapour phase

Rastogi 2001). More recently, Fick's second law has been used to describe moisture transfer in cellular solid foods (Voogt et al. 2011) and to describe water migration mechanisms in amorphous powder material and the related agglomeration propensity (Renzetti et al. 2012).

Fick's second law has also been used to predict moisture transfer in more complex systems, such as multidomain food products (or composite foods)—for example, breakfast cereal–raisin mixtures, ready-to-eat sandwiches, cereal-based pastry filled with a savory, moist filling, and so forth—in which moisture transfer occurs from the wet into the dry domain until thermodynamic equilibrium between the food components is reached (Guillard et al. 2003; Bourlieu et al. 2008; Roca et al. 2008). This moisture transfer affects the physical, sensory, and microbial qualities of the food and, especially, leads to the loss of texture for the dry cereal-based compartment in the common case of association of a crispy cereal texture with a moist filling. Modeling moisture transport within these products is of great interest in order to predict water distribution within the product and, thus, its shelf life. In the above-mentioned studies (Guillard et al. 2003; Bourlieu et al. 2008; Roca et al. 2008), Fick's second law succeeded in predicting water distribution in each domain of the composite foods, even in porous products, which were assimilated to continuous matter. Diffusivity was found to vary with moisture content (Guillard et al. 2003), which was successfully taken into account in the model, as well as various interfacial conditions such as nonperfect contact between the domains (Roca et al. 2008) and the presence of an edible barrier film at the interface between the domains (Guillard et al. 2003; Bourlieu et al. 2006). Such mathematical models, even if they did not represent the water transport mechanisms that prevail in the system, can be used as decision-making tools to design composite food with optimized shelf life (e.g., dimensions of the different domains, thickness of the edible barrier film, initial water activity of the wet compartment, etc.) (Roca et al. 2008).

2. Semi-theoretical Models

Semi-theoretical models offer a compromise between the theory and its ease of use (Fortes and Okos 1981). They are generally derived by simplifying the general series solution of Fick's second law. They are less time consuming than theoretical models in calculations. They are valid only within the temperature, RH, air flow velocity, and moisture content range for which they were developed, but the key point is that they do not need assumptions on food geometry, mass diffusivity, and conductivity, as compared with theoretical models such as those based on Fick's second law. As is shown in Table 2, semi-theoretical models developed for the drying of thin layers, such as the Henderson and Pabis model or the two-term model, relate moisture ratio of the food being studied to time through an exponential relation. The Henderson and Pabis model is the first term of a general series solution of Fick's second law (Henderson and Pabis 1961). This model was successfully used to model drying of corn (Henderson and Pabis 1961), wheat (Watson and Bhargava 1974), rough rice (Wang and Singh 1978), and mushrooms

(Gürtas 1994). The slope of the Henderson and Pabis model is under conditions related to the water effective diffusivity (Madamba et al. 1996). The two-term model is the first two terms of a general series solution of Fick's second law and has also been used to describe the drying of corn (Henderson 1974; Sharaf-Eldeen et al. 1980) and white beans and soybeans (Hutchinson and Otten 1983). However, it requires a constant product temperature and assumes a constant diffusivity. The Lewis model is a special case of the Henderson and Pabis model and was used to describe the drying of barley (Bruce 1985), wheat (O'Callaghan et al. 1971), and cashew nuts (Chakraverty 1984). These three semitheoretical models have been compared by Ozdemir and Devres (1999) for describing the isothermal roasting of hazelnuts. The two-term model obtained the highest correlation coefficient between experimental and predicted data, followed by the Henderson and Pabis model and, eventually, by the Lewis model. Nowadays, due to higher and higher computer capacities, these simplified models are less used.

3. Phenomenological Models

Unlike rigorous theoretical models, phenomenological models offer the possibility of modeling moisture transport phenomena without a priori hypotheses concerning the transport mechanisms that are involved, provided that the driving forces are known. A system under study is treated as a black box, and equations are written connecting the fluxes to the driving forces acting upon the system. Apparently, these equations are very similar to the ones describing rigorous steady-state equations in a Fickian diffusion. The difference in these phenomenological models lies in the fact that phenomenological coefficients are used instead of strict mechanistic coefficients characterizing moisture transport like diffusivities. Models based on irreversible thermodynamics theory belong to the phenomenological models. According to this theory, any flux is given by a linear relationship of all the driving forces of the system. The most common phenomenological model is the Luikov model (Table 2), which takes into account not only moisture and thermal diffusion, but also the cross-effects between thermal and moisture gradients (Soret and Dufour effects). Luikov's (1975) model has been widely used for heat and mass transfer in capillary porous products (Fortes and Okos 1981; Dantas et al. 2003) or in rough rice (Hussain et al. 1973), maize kernels (Neményi et al. 2000), wood and peanut pod (Kulasiri and Samarasinghe 1996). Another phenomenological model is to consider the difference in the chemical potential of water, instead of moisture content, as the driving force in the formulation of Fick's second law. By combining models based on irreversible thermodynamics and Fick's second law expressed with chemical potential, phenomenological coefficients (L_{ik}) can be determined.

The Stefan–Maxwell approach for mass transfer is a kind of model combining both a mechanistic and a thermodynamic character. It is thermodynamic in the sense that the chemical potential is used as the driving force. However, it is not a phenomenological model, because friction is assumed to be a transport mechanism. The main idea of the model is that the transport of a component, which is due to its potential, creates a force that is balanced by the friction with the surroundings. Friction is expressed in terms of velocity differences of a pair of components. Since fluxes are not explicitly treated contrary to Fick's laws, this model was progressively given up by experimentalists. Even so, the simplified version developed by Wesselingh and Krishna (1990), cited in Gekas (1992), transforming differentials in finite differences makes the Stefan–Maxwell approach very interesting. Whereas the Stefan–Maxwell approach is successfully used in material science for describing water transport, for example, through hydrogel (Hoch et al. 2003) or zeolite membranes (Gardner et al. 2002), it is rarely used in food science.

4. Empirical Models

Empirical models derive from a direct relationship between average moisture content and time. They are very useful if neither the mechanisms nor the fluxes or driving forces are identified clearly in a transport problem, which is easily the case in very complex systems. In general, the aim of such models is to establish a relationship between a dependent variable and a number of independent variables. They do not provide any information on the mechanisms occurring during the process, although they succeed in describing the moisture content evolution for the conditions of the experiment. Empirical models have been extensively used for describing the drying curve of various foodstuffs, especially when drying conditions are not well characterized or constant throughout the experiment. Examples of such models are the *Thompson* model and the *Wang and Singh* model (Table 2). The *Thompson* model was used to describe shelled corn drying for temperatures between 60° and 149 °C (Thompson et al. 1968), and the Wang and Singh model was used to describe drying of rough rice (Wang and Singh 1978). Because knowledge has considerably improved for knowing and controlling conditions of transfer (e.g., temperature, boundary conditions, etc.) in food processing, mechanistic approaches could now be use most of the time, and therefore, such empirical models have been given up.

In conclusion, the overall water transport in foods products is complex and results from different mechanisms, which cannot be easily distinguished. Molecular diffusion according to Fick's law is generally assessed as the main water transport mechanisms in food matrices (Eq. 6). The moisture diffusivity (D) in the foodstuff is, consequently, an apparent or effective diffusivity (D_{eff}) representative of the overall moisture transport. Even through this method is not perfectly sound theoretically, because Fick's law is extrapolated out of its application conditions, it is a very convenient and numerically accurate approach to describing moisture content transport during processing and storage, provided that effective diffusivity (D_{eff}) is well determined, as well as boundary conditions.

Determination of Water Diffusivity in Solid Products

To obtain an estimate of the water effective diffusivity for a given material in well-defined environmental conditions, two choices exist: (1) obtain values from the literature for the same or similar products or (2) measure these values. As regards water and solutes, (D_{eff}) databases exist, such as, for example, the one reported by Doulia et al. (2000). Nevertheless, the large variety of foodstuffs is not entirely represented in such databases, and, for a given material, (D_{eff}) is calculated in restrictive conditions of temperature and humidity range. In most cases, (D_{eff}) values cannot be extrapolated to different materials or in different conditions of temperature and humidity. Consequently, direct measurements of (D_{eff}) should be performed to determine precisely the moisture transport of the food.

Unfortunately, there is no standard method of diffusivity estimation. Contrary to water sorption isotherm, moisture diffusivity cannot be directly measured from experiments. Moisture transfer must be generated in the product under study and then monitored as a function of time and/or position in the food. Then experimental kinetics (e.g., average moisture content evolution with time or, for a given time, local water distribution as a function of position) must be fitted using a mathematical model (analytical or numerical solution of Fick's second law) in order to identify the diffusivity by using an optimization procedure. One of the common criteria used for optimization is the minimizing of the sum of squared error between experimental values and those predicted by the model.

a. Solving Fick's Second Law

In many cases, Fick's second law can be analytically solved if several assumptions are experimentally verified, such as, for example, uniform initial concentration, plane sheet or spherical geometry, monodirectional transfer, no external resistance to mass transfer (especially in the case of transfer between food and surrounding atmosphere), no shrinkage or swelling, and constant (D_{eff}). Such analytical solutions covering varying specimen geometry are found in the famous book of Crank (1975).

Unsteady mass transfer problems can also be solved numerically by transforming the partial differential equation of mass transfer into finite-difference equations in both space and time domains (Burden and Faires 1997). An advantage of such numerical solutions is that nonlinear terms of the partial differential equation can be solved covering presence of more complex initial and boundary conditions than with analytical solutions, including external mass transfer resistance, simultaneous mass transfer and chemical reaction, moving boundaries, and so forth. (D_{eff}) dependence on temperature and moisture content can also be taken into account through various multiparameter equations. Different schemes of discretization exist, such as the Crank–Nicolson one. Nowadays, due to the use of ODE solvers integrated in most commercial software (see, e.g., ode15 s or ode45 functions of MATLAB® software), only the spatial second-order derivatives in Eq. 6 are discretized using, for example, a three-point central difference formula.

Then the discretized form of Eq. 6 and its boundary and initial conditions is solved using an ODE routine that adjusts the size of the step of time for calculations according to the importance of moisture content variations. Such ODE solvers facilitate the solving of Fick's second-law-based system a lot and permit routine consideration of more complex assumptions for mass transfer (e.g., swelling or shrinkage of the product, diffusivity variations as a function of moisture content, etc.).

Both analytical and numerical solutions perform simulations of local moisture content in time and space domains in the food, which can allow calculation of the average moisture content evolution with time of the product being studied.

b. Identification of Diffusivity

Once a mathematical model simulating moisture transfer in the experiment is set up, diffusivity values or parameters of diffusivity law must be identified by comparison with experimental data. Either local or global moisture content evolution with time can yield an effective diffusivity or parameters of diffusivity law by using an optimization procedure. The most used criterion for optimization is the minimizing of root mean squared error between experimental and predicted data (Gill et al. 1981).

A variety of experiments are designed to suit the particular needs of the analytical or numerical solution chosen, including permeation methods, sorption or drying kinetics, and moisture distribution profiles. A detailed description of some of these methods was given by Zogzas et al. (1994) and is completed in the following paragraphs.

c. Experimental Setups for Diffusivity Determination

i. *Permeation Methods*

These methods were primarily developed for the evaluation of moisture diffusion through polymer membranes.

Steady state. A thin sheet of material is placed between two sources maintained at a constant concentration of diffusant (e.g., between two compartments that are maintained at different isothermal RH by means of a suitable buffer or saturated salt solutions). After a time period, the surfaces of the sheet come into equilibrium with the diffusant sources, thus developing a constant gradient of surface concentrations, leading to steady state conditions of diffusion. This state can be expressed, for a plane sheet, by the following equation:

$$J = D_{\text{eff}} \frac{(C_1 - C_2)}{L} \tag{7}$$

where C_1 and C_2 are concentrations (kg m^{-3}) in diffusing substance at each side of the layer and L is the thickness of the layer (m).

Diffusivity can be estimated by measuring the flux of the diffusant, with known surface concentrations and thickness of the material sheet, which can be done experimentally by successive weighing of the diffusion cells at different time intervals. Varzakas et al. (1999), Floros and Chinnan (1989), and Djelveh et al. (1988, 1989) successfully used this method for determining D_{eff} values of enzymes in polysulphone membranes, hydroxyl (OH−) through tomato skin, and NaCl through gels and meat products, respectively. This method was often applied to edible coatings (ethyl cellulose, zein, wheat gluten, etc.) (Bourlieu et al. 2009).

Time-lag method. This method is based on the time period prior to the establishment of the steady state diffusion. If, by some convenient means, one of the sheet surfaces is maintained at C_1 concentration while the other is maintained at zero concentration, after a theoretically infinite time period, a steady state condition of diffusion will be achieved. Assuming that the diffusivity is constant, that the sheet is initially completely free of diffusant, and that the diffusant is continually removed from the low-concentration side, the amount of diffusant that will permeate the sheet, when $t \to +\infty$, is given as a linear function of time by Eq. 8:

$$W_t = \frac{D_{eff} C_1}{L} \left(t - \frac{L^2}{6 D_{eff}} \right) \tag{8}$$

where W_t is the amount of diffusant (kg m^{-2}).

By plotting W_t against time, after a relatively large time interval, a straight line results, intercepting the t-axis at the quantity $(L^2/6D_{eff})$. From this intercept, D_{eff} may be deduced.

Although these two permeation methods seem to be simple in application, they raise experimental problems, including the difficulties of manufacturing a thin material sheet of constant thickness and homogeneous structure, the erroneous measurement of the flow rate of diffusion, and the swelling of the membrane material under experimental conditions. Therefore, this methodology is essentially used for the thin membranes of materials such as packaging film.

ii. *Water Sorption or Desorption (Drying) Kinetics*

During drying or desorption experiments, a geometrically defined sample (e.g., a slab of L thickness) is dried in a renewed air flow under conditions in which the partial pressure of water and temperature in the air are kept constant. Under these conditions, after an initial, very short period (constant rate period) during which the rate of drying is constant, there follows a so-called *falling rate period* (Fig. 5), in which the rate is assumed to be controlled by internal mass transfer resistance— that is, by the moisture diffusivity within the material. The weight loss or gain with time of the foodstuff is monitored. The usual practice has been to assume that the effective diffusion coefficient is indeed constant over the entire falling rate period or, at least, over significant portions of this period. Consequently, the simplified analytical solution of Fick's second law (Eq. 9) can be applied for modeling moisture transfer in a slab:

$$\frac{X - X_e}{X_0 - X_e} = \frac{8}{\pi^2} \exp \left[-\frac{\pi^2}{4} \frac{D_{eff}}{L^2} t \right] \tag{9}$$

Fig. 5 Typical drying curve
with (**a**) moisture content
evolution in the product,
(**b**) drying kinetic and
(**c**) temperature evolution in
the product (from Daudin
1983)

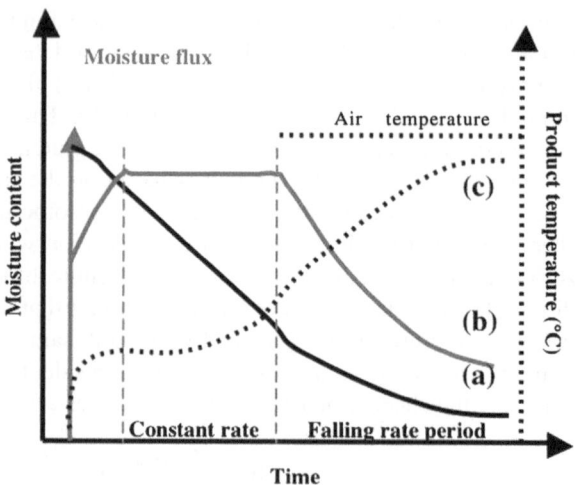

where X_e is the moisture content (kg m^{-3}) of the product in equilibrium with RH
in the air, X_0 is the initial moisture content (kg m^{-3}), and L is a characteristic
dimension (m) (e.g., slab thickness).

The drying process can then be represented by one or more straight lines, when
the quantity $\mathrm{Log}(X - X_e/X_0 - X_e)$ is plotted against time. Obviously, this method
is simple and has often been used for moisture diffusivity determination—for
example, by Biquet and Guilbert (1986) in agar gels, by Hebbar and Rastogi
(2001) in cashew kernel, by Lazarides et al. (1997) in fruit and vegetables tissues
undergoing osmotic processing, by Raghavan et al. (1995) in raisins, and so forth.
However, in Eq. 9, several assumptions are made, such as constant diffusivity,
surface area, and slab thickness, but also negligible external mass transfer resistance between air and sample. In reality, a modification of surface area due to
swelling, shrinkage, or porosity changes can occur in the food. Shrinkage can be
monitored, and a corrected geometry must be used in the calculations. However,
diffusivity dependence on moisture content cannot be represented using Eq. 9.
More appropriate alternative solutions of Fick's second law (numerical solution)
taking into account possible change in the dimensions of the sample, as well as the
particular boundary conditions and the diffusivity dependence on moisture content
and/or temperature, have thus been proposed and used. For example, Bonazzi et al.
(1997) developed a conventional diffusive model with variable diffusion coefficient, solved by finite difference calculations in a solid-related frame of coordi-
nates, in order to identify moisture diffusivity in gelatine slabs during drying. More
recently, Batista et al. (2007) also used a numerical scheme to obtain a diffusivity
law in thin chitosan layer, taking into account shrinkage. Such numerical solutions
are nowadays currently used in the field of drying technology, and a list of
examples could not be exhaustive.

Moisture changes in a foodstuff can also be measured in adsorption type under
controlled conditions of RH. The mass evolution with time is monitored up to

equilibrium. Time before reaching equilibrium varies as a function of sample geometries. The thinner the sample, the more rapidly the equilibrium will be achieved. Biquet and Labuza (1988) and Leslie et al. (1991) utilized this method with chocolate films and starch gels, respectively. The simplest way to control RH is to put the sample over saturated salt solutions (Labuza et al. 1976). The main advantage of this method lies in the less severe conditions imposed on the product, as compared with a conventional drying technique. As a consequence, the sample is subjected less to collapse or shrinkage. However, this methodology is time consuming (several days or even weeks before obtaining equilibrium). Consequently, as was explained above for the measurement of water sorption isotherms, microbial development can occur on the surface sample at high RH level and necessitate adding some antimicrobial additives.

Recently, the use of Cahn microbalance (or other, similar systems) to measure water vapor sorption kinetic in a thin piece of material (<30 mg) has permitted significantly reducing the time before equilibrium is reached and improving the study of water diffusivity in food and nonfood materials (Guillard et al. 2003c; Bourlieu et al. 2006; Oliver and Meinders 2011). The Cahn balance allows one to obtain a complete water adsorption or desorption kinetic for several successive RH levels in less than 4 days (depending on the material and temperature). From the transient state data obtained for each RH step studied, a diffusivity value can be identified. This methodology has the advantage of permitting one to observe diffusivity variations as a function of moisture content in the product without making assumptions a priori on the diffusivity law.

A significant problem is often encountered during both drying and sorption kinetics measurement by the formation of a boundary layer, offering a considerable external resistance to moisture transfer. This problem is faced again by stirring the surrounding fluids. However, stirring has proven inadequate, especially in cases where the internal resistance to mass diffusion is small enough. In that case, D_{eff} values characterize an overall mass transport representing both the internal diffusion and the external mass transfer at the interface. Since external resistance to mass transfer is difficult to determine experimentally, numerical solutions that take into account both the internal diffusion and the mass transfer coefficient at the interface can scarcely be applied for diffusivity identification. Nevertheless, in order to improve the accuracy of their moisture diffusivity values identified from water sorption kinetics measured in a Cahn microbalance, Roca et al. (2008) have determined the external mass transfer coefficient dependent of the geometrical and air flow characteristics encountered in the microbalance used by these authors (DVS from SMS, London) and routinely used this correction when identifying their diffusivity values.

iii. *Moisture Local Distribution Profiles*

In this method, the local concentration of the diffusant within the sample as a function of distance is determined for the case of one-dimensional diffusion. One of the best ways to determine the concentration distance profiles is by using a long cylindrical sample undergoing one-dimensional diffusion along its axis, in a sorption process. After a specified time interval, the cylinder is sliced in many equal parts, and the moisture content of each slice is gravimetrically or chemically analyzed. Another technique is to use two cylindrical samples of the same radius but different in diffusant concentration (Fig. 6). At time $t = 0$, the two cylinders are joined together, and after a specified time interval, the diffusant concentration profile along the axis can be determined by slicing and weighing the samples slices. Diffusivity is then evaluated by identification from the experimental distribution profiles, using either an analytical or a numerical solution of Fick's second law with appropriate boundary conditions.

Contrary to sorption or drying kinetics, where the global loss or gain of moisture as a function of time is measured, the moisture distribution profile method directly provides spatial information about moisture diffusion, which can be compared with the local moisture contents predicted by the model. That means that diffusivity can be identified from a set of data obtained for only one diffusion time. Therefore, the diffusivity can be obtained more rapidly with a local distribution profile than with a global profile, provided that diffusivity is high enough to permit detection and quantification of water content in a thin slab of material.

Warin et al. (1997), Motarjemi (1988), Karathanos and Kostaropoulos (1995), and Litchfield and Okos (1992) have successfully used this method for measuring sucrose diffusivity within agar gels, moisture distribution profiles in meat, moisture distribution in dough/raisin systems, and water diffusivity in spaghetti, respectively. Recently, this approach was revisited by Guillard et al. (2003a) to identify moisture diffusivity in sponge cake. In their study, a cylinder of sponge cake was put in contact with a high moisture content compartment (agar gel). This

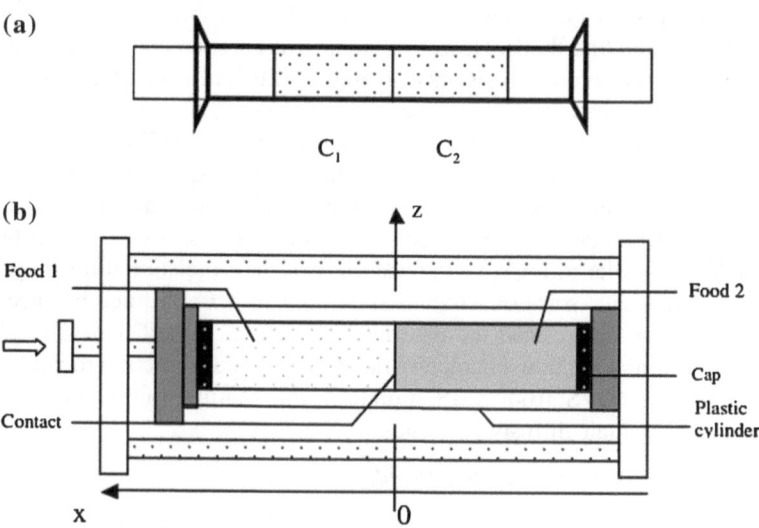

Fig. 6 Different experimental systems 'two-phases in contact » permitting one to obtain experiment data for diffusivity identification : (**a**) system based on two syringes used by Motarjemi (1988) to measure distribution profiles in meat, (**b**) system of Rougier (2007)

approach was also used by Boudhrioua et al. (2003) to estimate water diffusivity in gelatin–starch gels and by Roca et al. (2006) in a cereal-based product.

However, the moisture loss during slicing of material due to diffusion or evaporation could lead to nonnegligible errors. To avoid diffusion during slicing, Litchield and Okos (1992) have sliced frozen spaghetti. It is also practically difficult to obtain slices strictly perpendicular to the moisture transfer and, consequently, to have a good repeatability. And finally, the major problem of this technique is to be definitively invasive and destructive and, thus, time consuming. Therefore, several authors have tried to use a noninvasive methodology, such as magnetic resonance imaging (MRI), to obtain a local distribution profile.

MRI is an established technique for non-invasive measurements of moisture and moisture migration in food systems. Conventional MRI techniques are able to produce moisture maps in a very small parallelepiped volume of about 1 mm^3 with a high degree of resolution. Up to now, MRI has been extensively used in the medical field and has recently been paid more attention as a research tool for studying the drying or rehydration kinetics of food products (Ziegler et al. 2003; Wang et al. 2000; Ruiz-Cabrera 1999; Chen et al. 1997; Schrader and Litchfield 1992). For example, Wang et al. (2000) used MRI to monitor the change in water content in bread/ barbeque-chicken bilayer or bread/cheese sandwich systems. Hwang et al. (2009) applied MRI to examine water distribution and migration in rice kernels.

Another nondestructive method consists of measuring the absorption of gamma radiation going through a material. The absorption rate can be related to the moisture content of the sample part passed through by the radiation (Chiang and Petersen 1987).

Factors Affecting the Moisture Diffusivity Values

The main external parameters controlling the rate of water diffusion are temperature (Arrhenius relationship) and RH. Internal parameters modulating water diffusion are the structure of the food product and, particularly, the porous structure, which induces gaseous diffusion, on the one hand, and the local viscosity of the adsorbed aqueous phase, which, on the other hand, impacts liquid phase diffusion. Variations of apparent water diffusivity resulting from modifications of these parameters (temperature, moisture content, and/or porosity) have been reported for a large range of products.

a. Diffusivity Sensitivity to Moisture Content

The sensitivity of effective diffusivity (D_{eff}) to RH and, thus, water content is well documented and widely accepted. Figure 7 shows this relation for a variety of food components. Numerous studies have shown that effective diffusivity varied with moisture content, such as, for example, in agar gels (Biquet and Guilbert 1986), in corn-based extruded pasta (Andrieu et al. 1988), in bread, biscuits, and muffins (Tong and Lund 1990), and in sponge cake (Guillard et al. 2003a). Contrary to the typical D_{eff} increase usually observed as a function of moisture content, Karathanos et al. (1990) and Marousis et al. (1989) reported that, in porous starch mixtures, D_{eff} can increase with moisture content up to a limit value of 10 % (dry basis) and, then, decrease with moisture content and eventually becomes constant at sufficiently high moistures. This bell-like curve for D_{eff} versus moisture content or water activity has recently been observed in hydrophilic materials such as wheat gluten (Angellier-Coussy et al. 2011), starch (Chivrac et al. 2010) or sponge cake and other dry biscuits (Guillard et al. 2003c; Guillard et al. 2004b) (see Fig. 7). This particular behavior was related to change in the structure of the material.

Concerning these structural evolutions with a_w and their impacts on D_{eff}, three kinds of products can be distinguished:

- *Dense, hydrophilic products*, such as starch or wheat gluten edible films. In these materials D_{eff} first remained low until an a_w of 0.4, then it increased up to an a_w of 0.7–0.8, and finally decreased (Fig. 7). The low and constant D_{eff} value obtained for $a_w < 0.4$, followed by an increase in D_{eff} for $0.4 < a_w < 0.7$–0.8, could be related to changes in the mobility of polymer chains (glass transition) due to the plasticizing effect of water. In the high a_w range, the decrease of the apparent diffusivity in the rubbery wheat gluten film is attributed to a water clustering phenomenon, which would lead to an immobilization of sorbed water molecules and an apparent slower water movement (Gouanve et al. 2007). A decrease in D_{eff} with increasing moisture content was also observed in HPMC films by Bilbao-Sainz et al. (2011). They also interpreted the decrease in D_{eff} as a clustering of water subsequent to the increase in

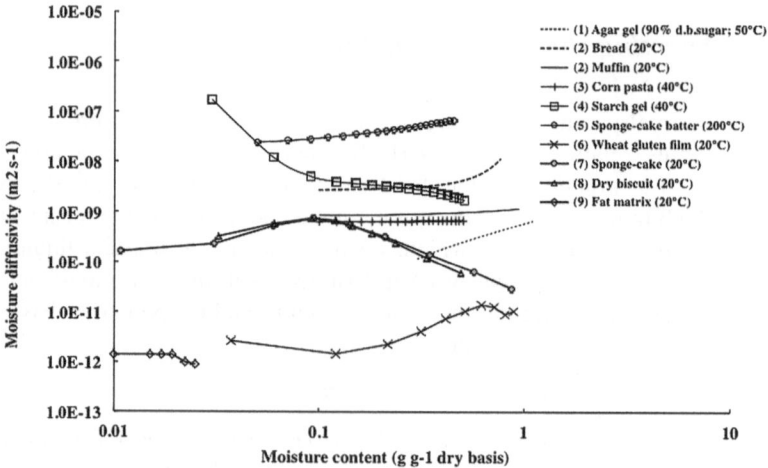

Fig. 7 Diffusivity of water as a function of moisture content in some food materials. (1) Biquet and Guilbert (1986), (2) Tong and Lund (1990), (3) Andrieu et al. (1988), (4) Marousis et al. (1991), (5) Baik and Marcotte (2003), (6) Angellier-Coussy et al. (2011), (7) Guillard et al. (2003a); (8) Guillard et al. (2004b) and (9) Bourlieu et al. (2006)

water content resulting in "free water." These free molecules, without inter-action with the matrix, aggregate to form di-, tri-, and tetra-mer clusters. These clusters present higher molecular volume than do monomers; thus, they diffuse more slowly and induce a drop in D_{eff}.

- *Porous products*, such as sponge cake. In these materials, through water adsorption, the matrix swells, leading to a decrease in the porosity of the product. As a consequence, water liquid diffusion in the solid matrix would predominate at this stage, as compared with the water vapor diffusion in the open pores. Since liquid diffusion of water (about 10^{-9} m^{2}s^1) is slower than vapor diffusion of water (about 10^{-5} m^2/s^1), the resulting effective water dif-fusivity decreased with the increasing moisture content of the sponge cake. This impact of swelling and decrease in porosity was pointed out by Environmental SEM by Guillard et al. (2003c) (see Fig. 8). Through water adsorption in sponge cake, a swollen matrix was formed, and the apparent degree of porosity decreased strongly near saturation, leading to a predominant water liquid dif-fusion. Then, near saturation, water diffusivity of the sponge cake was quite constant, because most of the porous structure had collapsed, and water trans-port occurred mainly in the liquid phase within the swollen matrix.
- Conversely, in *dense highly hydrophobic* materials, such as lipid-based edible barrier films (such as WAX/ACETOGLYCERIDES composite materials), sorbing a very small amount of water (GAB parameters of $Q_{mono} = 0.0085$, $C = 1.2351$, $K = 0.7893$), a quasi-constant D_{eff} ($\sim 0.05 \times 10^{10}$ m^2 s^{-1}) was reported (Bourlieu et al. 2008). This low D_{eff} was coherent with the hydrophobic partially crystalline nature of the material (solid fat content of 88 % at the temperature considered). At high a_w, the sorption of water did not plasticize this inert matrix or modify its barrier properties.

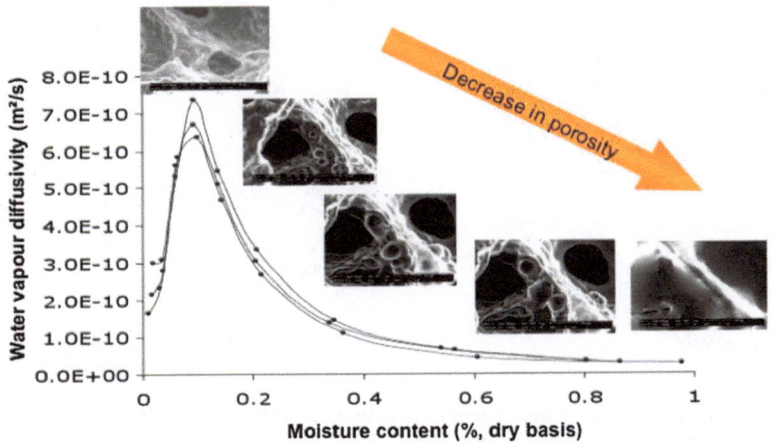

Fig. 8 Evolution of water vapour diffusivity in sponge cake as a function of moisture content and subsequent change in porosity as observed by ESEM—adapted from Guillard et al. (2003a, d)

It is clear from these results that structural parameters (e.g., porosity change, glass transition, etc.) are more relevant parameters for describing water diffusivity change in food and nonfood products than is an increase or decrease in product moisture content. Nevertheless, up to now, the variations of water diffusivity were scarcely quantitatively related to structural changes, and the classical relationship $D_{\text{eff}} = f(Q_w)$ is still used in food science.

b. Typical Values of Water Vapor Diffusivity in Foodstuffs

According to the various structures of food matrices, typical values of effective diffusivity varied from 1×10^{-9} to $1 \times 10^{-12}\,\text{m}^2\,\text{s}^{-1}$, with higher values for porous products such as cereal-based foods (Guillard et al. 2003a) and lower values for dense materials such as raisins or dates (Gencturk et al. 1986; Tütüncü and Labuza 1996) (Fig. 9). Extreme range of values of diffusion reported for the sake of comparison reaches $2.4 \times 10^{-5}\,\text{m}^2\,\text{s}^{-1}$ for water vapor in air, whereas self-diffusion coefficient of liquid water is $2 \times 10^{-9}\,\text{m}^2\,\text{s}^{-1}$ and lower limits reported in glassy materials average $2 \times 10^{-14}\,\text{m}^2\,\text{s}^{-1}$ (Gekas 1992). Moisture diffusivity is very sensitive to the variations in composition and structure of the examined materials. But even for the same food material, variations in diffusivity values are generally encountered between two investigations, proving that diffusivity is also very sensitive to the particular experimental and mathematical methods used for its determination.

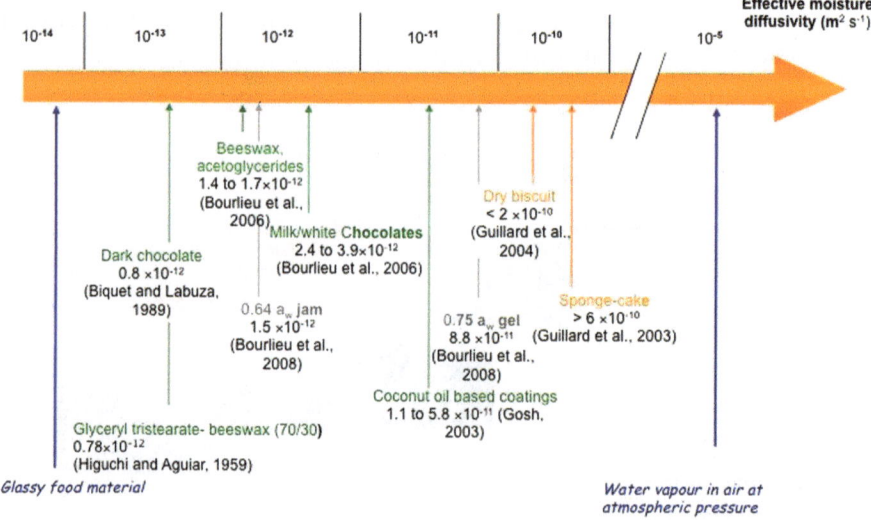

Fig. 9 Range of values for water vapour diffusivity in foodstuffs and edible coatings (data from Biquet and Labuza 1988; Bourlieu et al. 2006, 2008; Ghosh 2003; Guillard et al. 2003, 2004; Higuchi and Aguiar 1959)

c. Validity and Accuracy of Identified Diffusivity Values

No standard method of diffusivity estimation exists. As regards to the economical importance of drying, many investigators have been interested in modeling drying kinetics in order to optimize processes and, thus, have determined moisture diffusivity in food matrices subject to drying. Whatever the experimental method used, the same analytical treatment of experimental data is usually performed. Moisture diffusivity is identified by minimizing error between experimental and predicted moisture contents obtained by experimental monitoring of mass change during drying, desorption, or sorption experiment. When D_{eff} is suspected to vary with moisture content, which is almost always verified, a diffusivity law as a function of moisture content is chosen a priori, and the parameters of this law are identified in the same manner as the constant value.

Zogzas et al. (1996) reviewed the different D_{eff} laws as a function of moisture content encountered in the literature. Contrary to D_{eff} variations with temperature, which are almost always represented through an Arrhenius equation, diffusivity dependence on moisture content, whose variety of behaviors have been previously displayed in Fig. 7, has been represented by various empirical relations relying on numerous parameters (from 3 to 5) (Table 3).

More complex relations are provided when simultaneous temperature and moisture content, but also porosity, dependence of D_{eff} is taken into account as listed in Table 3. From this complexity of relations but, also, of methods used for generating experimental data and mathematical solutions chosen for analytical treatment of data, large variations in D_{eff} values are observed even for the same material, such as in corn, where it was found that D_{eff} varied from 5×10^{-12} to 8×10^{-7} m^2 s^{-1} for the same moisture content range.

Water Vapor Permeability

The water vapor permeability (WVP) of a film is a steady state property that describes the rate of water transfer through the film and summarizes the phase of adsorption, water vapor diffusion in the film matrix, and desorption on the other side of the film. This rate is obtained for a given RH difference, which is applied on each side of the film usually using a permeation cell that contains a desiccant or saturated solution and is stored in a desiccator at a given RH and is hermetically sealed with the barrier film (American ASTM E96-80; ASTM 1980). Comprehensive tables of WVP values have recently been proposed by Morillon et al. (2002), Wu et al. (2002), and Bourlieu et al. (2009b). Examples of values taken from Bourlieu et al. (2009) summarize the range of moisture WVP values encountered in food coating, in comparison with polymeric packaging, at 20 °C are presented in Table 4.

Table 3 Empirical parametric models expressing diffusivity as a function of temperature and moisture content for various food materials (X g/g, T degree K)

Num. of par	Material	Parametric model	Reference
2		$D(X) = a_0 + a_1 X$	
	Swordfish	$a_0 = 9 \times 10^{-11}$, $a_1 = 1.6 \times 10^{-10}$ $1.0 \leq X \leq 5.0$, $313 \leq T \leq 328$	Del Valle and Nickerson (1968)
	Cereals	$a_0 = 1.35 \times 10^{-2}$, $a_1 = 4.2 \times 10^{-4}$ $0.1 \leq X \leq 0.3$, $T = 330$	Whitaker et al. (1969)
		$D(X) = D_0 X^\alpha$	
	Sponge cake	$D_0 = 2.3 \times 10^{-11}$, $\alpha = -1.6$ $0.20 \leq X \leq 0.70$ (wet basis), $T = 293$	Guillard et al. (2003c)
3		$D(X,t) = a_0 \exp(a_1 X) \exp\left(-\frac{a_2}{T}\right)$	
	Corn	$a_0 = 7.817 \times 10^{-5}$ (or 4.067×10^{-5}), $a_1 = 5.5$, $a_2 = 4850$ $0.05 \leq X \leq 0.4$, $283 \leq T \leq 300$	Parti and Dugmanics (1990)
	Carrot	$a_0 = 1.053 \times 10^{-4}$, $a_1 = 0.059$, $a_2 = 3460$ $0.0 \leq X \leq 5.0$, $303 \leq T \leq 343$	Mulet (1994)
	Starch (granular)	$a_0 = 5.321 \times 10^{-6}$, $a_1 = -1.511$, $a_2 = 2848.5$ $0.11 \leq X \leq 0.5$, $298 \leq T \leq 413$ at 1 atm pressure	Karathanos et al. (1994)
		$D(X,T) = a_0 \exp\left(-\frac{a_1}{X}\right) \exp\left(-\frac{a_2}{T}\right)$	
	Onion	$a_0 = 3.72$, $a_1 = 8.63 \times 10^{-2}$, $a_2 = 7110$ $0.05 \leq X \leq 18.7$, $333 \leq T \leq 353$	Kiranoudis et al. (1992)
	Potato	$a_0 = 1.29 \times 10^{-6}$, $a_1 = 7.25 \times 10^{-2}$, $a_2 = 2044$ $0.03 \leq X \leq 5.0$, $333 \leq T \leq 373$	Kiranoudis et al. (1995)
	Carrot	$a_0 = 2.71 \times 10^{-7}$, $a_1 = 7.44 \times 10^{-2}$, $a_2 = 1527$ $0.03 \leq X \leq 5.0$, $333 \leq T \leq 373$	Kiranoudis et al. (1995)
		$D(X,T) = a_0 [1 - \exp(-a_1 X)] \exp\left(-\frac{a_2}{T}\right)$	
	Starch gel (amioca)	$a_0 = 4.8 \times 10^{-6}$, $a_1 = 2.8$, $a_2 = 3668.4$; $00.1 \leq X \leq 1.0$, $T = 333$ $a_0 = 4.8 \times 10^{-6}$, $a_1 = 0.9$, $a_2 = 3668.4$; $00.1 \leq X \leq 1.0$, $T = 373$	Vagenas and Karathanos (1993)
		$D(X,T) = (a_0 + a_1 T)\left(\frac{X_0}{1+X_0} - \frac{X}{1+X}\right)^{a_2}$	
	Starch gel	$a_0 = 12.96 \times 10^{-11}$, $a_1 = 2.72 \times 10^{-11}$, $a_2 = 0.02$ $0.2 \leq X \leq 1.7$, $303 \leq T \leq 323$, pressure: $P = 1$ bar	Buvanasundaram et al. (1994)
		$D(X,T) = a_0 \exp\left[-\frac{a_1}{T} + a_2 X\right]$	
	Baking cake	$a_0 = 9.75$, $a_1 = 7689.9$, $a_2 = -0.0858$ $0.369 \leq X \leq 0.707$, $306 \leq T \leq 395$	Baik and Marcotte (2002)
4		$D(X,T) = (a_0 + a_1 X + a_2 X^2) \exp\left(-\frac{a_3}{T}\right)$	
	Pasta (corn)	$a_0 = 1.15 \times 10^{-3}$, $a_1 = 1.67 \times 10^{-2}$, $a_2 = 7.8 \times 10^{-2}$, $a_3 = 5840$ $0.1 \leq X \leq 0.5$, $313 \leq T \leq 353$	Andrieu et al. (1988)

(continued)

Table 3 (continued)

Num. of par	Material	Parametric model	Reference
	Apple	$D(X,T) = a_0 \exp(a_1 X) \exp\left(-\frac{a_2 X + a_3}{T}\right)$ $a_0 = 2.75 \times 10^{-3}$, $a_1 = 12.97$, $a_2 = 4216$, $a_3 = 5267$ $0.04 \leq X \leq 1.3$, $288 \leq T \leq 318$	Singh et al. (1984)
	Corn (shelled)	$D(X,T) = a_0 \exp[(a_1 T - a_2)X] \exp\left(-\frac{a_3}{T}\right)$ $a_0 = 4.254 \times 10^{-8}$, $a_1 = 4.5 \times 10^{-2}$, $a_2 = 5.5$, $a_3 = 2513$ $0.05 \leq X \leq 0.35$, $323 \leq T \leq 343$	Chu and Hustrulid (1968)
	Corn	$a_0 = 2.54 \times 10^{-8}$, $a_1 = 1.2343 \times 10^{-1}$, $a_2 = 45.47$, $a_3 = 1183.3$ $0.1 \leq X \leq 0.4$, $293 \leq T \leq 338$	Ulku and Uckan (1986)
	Corn	$a_0 = 1.905 \times 10^{-5} X_0$, $a_1 = 1.83 \times 10^{-2}$, $a_2 = 2.37$, $a_3 = 2153$ $0.1 \leq X \leq 0.45$, $323 \leq T \leq 393$	Mourad et al. (1994)
	Rice	$a_0 = 2.744 \times 10^{-6}$ $a_1 = 1.589 \times 10^{-3}$, $a_2 = 0.379$, $a_3 = 4294.8$ $0.26 \leq X \leq 0.32$, $321.8 \leq T \leq 355.2$	
5		$D(X,T) = a_0 \exp\left(\sum_{i=1}^{3} a_i X^i\right) \exp\left(-\frac{a_4}{T}\right)$	
	Bread	$a_0 = 2.8945$, $a_1 = 1.26$, $a_2 = -2.76$, $a_3 = 4.96$, $a_4 = 6117.4$ $0.1 \leq X \leq 0.75$, $293 \leq T \leq 373$	Tong and Lund (1990)
	Biscuit	$a_0 = 0.92114$, $a_1 = 0.45$, $a_2 = 0$, $a_3 = 0$, $a_4 = 6104.5$ $0.1 \leq X \leq 0.6$, $293 \leq T \leq 373$	Tong and Lund (1990)
	Muffin	$a_0 = 6.16729$, $a_1 = 0.39$, $a_2 = 0$, $a_3 = 0$, $a_4 = 6664.0$ $0.1 \leq X \leq 0.95$, $293 \leq T \leq 373$	Tong and Lund (1990)
	Dry biscuit	$D(X) = D_0 \sum_{i=1}^{n} \alpha_i X^i$ $D_0 = 3.06 \times 10^{-10}$, $a_1 = -628.30$, $a_2 = 733.33$, $a_3 = -287.94$, $a_4 = 36.96$ $0 \leq X \leq 0.30$ (wet basis), $T = 293$	Guillard et al. (2004b)
	Pasta (dur. semolina)	$D(X,T) = a_0[1 - \exp(-a_1 X^{a_2}) + X^{a_3}] \exp\left(-\frac{a_4}{T}\right)$ $a_0 = 2.392 \times 10^{-7}$, $a_1 = 7.9082 \times 10^{-14}$, $a_2 = 15.7$, $a_3 = 0.6859$, $a_4 = 3156.3$ $0.015 \leq X \leq 0.26$, $313 \leq T \leq 353$	Litchfield and Okos (1992)
	Starch gel	$D(X,T) = \left[a_0 + a_1 X^{-a_2} + a_3 \frac{\varepsilon^3}{(1-\varepsilon)^2}\right] \exp\left(-\frac{a_4}{T}\right)$ $a_0 = 4.84 \times 10^{-7}$, $a_1 = 5.735 \times 10^{-11}$, $a_2 = 4.337$, $a_3 = 3.42 \times 10^{-6}$, $a_4 = 2264$ $0.03 \leq X \leq 0.5$, $313 \leq T \leq 373$, pressure : $P = 1$ bar $\varepsilon = (0.6 + 0.0643X$ $-0.37X^2) \exp(0.0077 - 0.113X \times P)$	Marousis et al. (1991)

Table 4 Comparison of moisture water vapor permeability (WVP; 20 °C) ranges for edible coatings in comparison with most commonly used synthetic packaging films

Film or coating	WVP (10^{-11} g m^{-1} s^{-1} Pa^{-1})	RH difference (%)	Thickness (mm)	Temperature (°C)	Reference
Synthetic films					
Aluminium	0.0005	0–98	(–)	38	(Myers et al. 1961)
HPDE	0.002	0–100	0.019	27.6	(Shellhammer et al. 1997)
PP	0.010		0.025	25	
PVC	0.041	0–100	0.012	27.6	
Lipid films	0.03–1.0			20	
Waxes					
Candelilla wax	0.014	0–100	0.14	24.9	(Shellhammer et al. 1997)
Paraffin	0.023		0.66	25	(Lovegren and Feuge 1954)
Beeswax	0.103		0.14	25.9	(Shellhammer et al. 1997)
Carnauba wax	0.114		0.130	27.5	
Fatty acids					
Capric acid	0.38	12–56			(Koelsch and Labuza 1992)
Palmitic acid	0.65				
Stearic acid	0.22				
Triglycerides & derivatives					
Tripalmitin	0.225	0–100	0.130	27.5	(Shellhammer and Krochta 1997a)
Triolein	12.100	22.84		25	(Quezada Gallo 1999)
Anhydrous milkfat	1.028	0–100	0.130	24.9	(Shellhammer et al. 1997)
Hydrogenated peanut oil	3.863		3.39	25	(Lovegren and Feuge 1954)
Glyceryl monostearate	0.77		1.026	24.3	(Higuchi and Aguiar 1959)
Acetomonopalmitine	11.35		0.200	20	(Bourlieu et al. 2006)
Tempered cocoa butter	26.8–61.2		1.59–2.92	26.7	(Landmann et al. 1960)
Milk chocolate	88.99		0.68	20	(Bourlieu et al. 2006)
Protein films	1.0–100.0	0–100		20	

(continued)

Table 4 (continued)

Film or coating	WVP(10^{-11} g m^{-1} s^{-1} Pa^{-1})	RH difference (%)	Thickness(mm)	Temperature (°C)	Reference
Wheat Gluten		12.97	0.053	20	(Guillard et al. 2003a)
Soya	281.18	50–66	0.072	25	(Rhim et al. 2002)
Corn zein	11.6	0–85	0.12–0.33	21	(Park and Chinnan 1995)
Polysaccharide films	1.0–10.0	0–100		20	
Cellulose derivatives	9.2–11.0	0–85	0.04–0.07	21	(Park and Chinnan 1995)
Starch	25–78	11–100	0.005–0180	30	(Bertuzzi et al.)

From a theoretical point of view, the WVP through a barrier is determined combining Fick's first law of diffusion (Eq. 7) with Henry's law of solubility (Eq. 10) as expressed in Eq. 11 (Park 1986; ASTM 1980):

$$Q = S \cdot p \tag{10}$$

where Q is the concentration of the permeate (mol m^{-3}), S the solubility coefficient defined as the maximum mass of the migrating molecule that dissolves in a unit volume of the material at equilibrium (mol m^{-3} Pa^{-1}), and p the permeate partial pressure in the adjacent air (Pa):

$$\text{WVP} = D_{\text{eff}} \cdot S = \frac{J \cdot e}{A \cdot \Delta p \cdot M} \tag{11}$$

where WVP is the water vapor permeability (mol m^{-1} s^{-1} Pa^{-1}), A the surface of barrier (m^2), M the water molar weight (g mol^{-1}), and e (m) the barrier thickness.

As is expressed in Eq. 11, WVP is the resultant of two parameters: a thermodynamic parameter, solubility, which depends on the compatibility between the penetrant molecule at equilibrium and the material the penetrant is migrating through, and a nonthermodynamic kinetic parameter D_{eff}, indicating water mobility in the material and highly influenced by the structural and morphological characteristics of the material. The range of validity of Eq. 11 is limited to thin, nonporous, hydrophobic film, with low water vapor solubility, nonswelling, and constant permeation rate over time, whereas in most other cases (hydrophilic material that interacts with water), a given value of WVP cannot be considered as an inherent property of the film, since its value will be influenced by extrinsic parameters of the WVP test such as the difference of RH, the temperature of the test, and the thickness of the barrier (McHugh et al. 1993; McHugh and Krochta 1994; Morillon et al. 2002).

The WVP is often preferred over diffusivity value to characterize the moisture barrier property of edible films and coatings. Even if this parameter represents an overall moisture transport mixing both sorption and diffusion phenomena, it is convenient to classify and compare the ability of edible films to prevent moisture loss or gain when applied directly onto a food surface or when used at the interface between two food compartments of contrasted initial water activities. In the last case, the edible film aims at preventing moisture exchange between the high a_w compartment to the one with a lower a_w. Usually, the low a_w compartment is a cereal-based compartment such as, for example, a crispy biscuit or wafer (Guillard et al. 2004a; Bourlieu et al. 2006) associated to a moist filling such as jam, pastry cream, and so forth. Edible barrier films, almost always lipid-based films, are very interesting solutions to slow down moisture transfer between food compartments and to increase the shelf life of the overall product, limiting texture loss (Roca et al. 2008).

However, WVP is usually measured for a given RH difference, often 0–100 % RH, which is rarely representative of the conditions that prevail in multicompartment foods. For example, classical association of a crispy wafer with a jam provided a difference of a_w around 0.14–0.50 (Bourlieu et al. 2006) or of a sponge cake with a moist filling an a_w difference of 0.84–0.95 (Guillard et al. 2003a). If WVP remains quite constant on the entire RH range for pure hydrophobic films, it is not the case for materials containing hydrophilic components. A typical example of such material very often used in food industry as edible film and coating is chocolate. For an $a_w > 0.80$, sugar particles contained in the continuous fat layer of chocolate begin to sorb a lot of water, to dissolve, and to swell. This is revealed by a sharp increase of the sorption isotherm curve of chocolate (Guillard et al. 2003d) and a nonlinearity of WVP evolution with RH difference for RH above 80 %. Consequently, if chocolate layers are good moisture barrier coating a low a_w (below 0.80), they are definitively inefficient for high a_w. Therefore, to be more predictive of the moisture barrier efficiency, some authors (Guillard et al. 2003d; Bourlieu et al. 2006) have used diffusivity coefficient and water vapor sorption isotherm instead of WVP as input parameters in Fick's second law for predicting moisture transfer through edible film placed at the interface between two domains of a composite food. They demonstrated that this approach was much more accurate in classifying moisture barrier efficiency of edible films and coatings.

Relationship Between Multiscale Food Structure and Moisture Transfer Properties

The structure of the food materials has a major impact on the diffusion rate and, consequently, on the value of effective diffusivity or water vapor permeability. On the contrary, water sorption at equilibrium is less or not affected by the structure. Among structure type, porosity particularly affects the diffusion rate. In a porous domain, molecules are allowed to transfer more quickly because of several mechanisms of moisture transfer (e.g., capillary action), vapor diffusion in the pores, and so forth, along with liquid diffusion. In low-porosity materials, liquid diffusion is the main means of moisture transport. The smaller the pores' size in the matrix of the food domain, the slower the moisture migration. In addition, membranes, crystals, and lipids all contribute as barriers to moisture migration. In such dense materials, elements of structure affecting the diffusion rate are in the nano- or microscale, whereas for porous materials, structural elements at the micro- and macroscales are more predominant.

In the following, a multiscale analysis of structural elements affecting water diffusion and water vapor permeability in foodstuffs will be performed.

Molecular Scale

Semicrystalline Materials

For dense materials such as edible coating, the main structural factor affecting moisture transfer is the tightness of the molecular structure. This influence is easy to understand considering the diffusion process "as a series of activated jumps from a vaguely defined cavity to another," as it has been described by Rogers (1985). Qualitatively, the diffusion rate rises with the increase of the number or size of cavities caused, for instance, by the presence of substances such as plasticizers. Indeed, for polymer-based films for instance, plasticizers weaken intermolecular forces between adjacent polymer chains and create voids, which results in a lower water resistance (Bourlieu 2007; Gontard et al. 1992; Navarro-Tarazaga et al. 2008). On the other hand, structural entities such as

V. Guillard et al., *Food Structure and Moisture Transfer*, SpringerBriefs in Food, Health, and Nutrition, DOI: 10.1007/978-1-4614-6342-9_3, © The Author(s) 2013

crosslinking or degree of crystallinity decrease the number of cavities and, thereby, the diffusion rate (Chao and Rizvi 1988).

For semicrystalline materials, such as lipidic substances, some modeling approaches were carried out to tentatively relate the crystals' domain size (influenced by crystallization conditions) and solid fat content (SFC) of a fat material to its WVP. Martini et al. (2006) proposed a mechanistic model to relate WVP of an edible lipid-base coating to its structural characteristic:

$$\frac{WVP}{WVP_{max}} = e^{-\alpha(\varepsilon/d)\phi} \tag{1}$$

where WVP is the water vapor permeability of the fat material (mol m m^{-1} s^{-1} Pa^{-1}), WVP$_{max}$ is the water vapor permeability of the oil in the absence of solids for $\varphi = 0$, φ is the volume fraction of solid (or SFC in %), α is the ratio of the crystal domain size ξ (A°) to the lamellar spacing d (A°), a parameter related to the tortuosity of the diffusional path through the crystal network.

Some other mathematical models, taken from the polymer science, help to understand the sorption behavior in a semicrystalline material. For instance in semicrystalline polymer, crystalline regions are characterized by extremely low moisture sorption (limited to the surface of the crystals) and extremely low D_{eff} values due to fixed configuration of molecules. The effect of the crystallinity degree on solubility may be described as a simple two-phase model (Bove et al. 1996):

$$S = S_a \varphi_a \tag{2}$$

where S_a is the solubility in the amorphous phase and φ_a is the amorphous phase volume fraction.

This expression, although widely used, must be handled with care, since it is only an approximation and does not account for the changes in the amorphous phase due to varying levels of crystallinity or, at high penetrant concentrations, for the crystallization induced by the sorbed molecule (Serad et al. 2001; Dhoot 2004). If Eq. 2 is well known and applied in polymer science, its use in food science for predicting water vapor sorption is scarce, probably because semicrystalline food products are often based either on (1) amorphous phase/hydrophilic crystals or (2) liquid phase/hydrophobic crystals with regard to lipids. For the first case of food matrices, which include hydrophilic crystals that are characterized by a type III isotherm according to the Brunauer classification, presenting a sharp upturn in moisture sorption at a a_w corresponding to crystal dissolution threshold, Eq. 2 cannot be applied. Considering now highly hydrophobic food matrices based on crystalline phase dispersed in liquid phase, Eq. 2 can be applied, replacing φ_a by the liquid phase volume fraction. In practice, the number of moisture sorption isotherms reported for semicrystalline hydrophobic food matrices is limited (Guillard et al. 2004b, c; Bourlieu et al. 2006, 2008, 2009a). Moreover, in those studies, this model was not applied.

Amorphous Materials (The Free Volume Theory)

In amorphous regions, where changes in polymer chain mobility are responsible for diffusivity variations, the free volume theory is usually considered to explain the diffusivity variations. In the free volume theory, the amorphous fraction of a semicrystalline polymer may be regarded as a network of polymer chains containing free spaces between them with different sizes and shapes. Actually, due to packing inefficiencies and polymer chain molecular motion, some of the volume in the polymer matrix is empty or free. This so-called free volume is redistributed continuously. The thermal agitation makes the polymer molecules move so that the distribution and the location of "holes" changes. Inside these local gaps, the diffusing molecules vibrate at much higher frequencies than the polymer chain motion. According to the free volume theory, a molecule will diffuse "jumping" through these holes only if the hole reaches a volume equal to or larger than the volume of the diffusing molecule (Wesselingh and Krishna 2006). The diffusing rate of a molecule will be proportional to the probability of finding such a neighboring hole. This probability is a negative exponential function of the molecular volume of the molecule and the available space in the polymer (Duda and Zielinsky 1996). Diffusivity may be expressed, then, as a function of the volume of the diffusing molecule and as the free volume present in the polymer as follows:

$$D = D_0 \exp\left(\frac{-E_D}{RT}\right) \exp\left(\frac{-\gamma V_1^*}{V_{FH}}\right) \qquad (3)$$

where V_1^* is the critical molar free volume required for a jumping unit of the diffusing molecule to migrate, V_{FH} is the free volume per mole of all individual jumping units in the solution, and D_0 is a temperature-independent constant. The first exponential is an Arrhenius-like equation.

There have been a number of corrections to this equation that are beyond the scope of this document. In any case, the core of the theory remains unchanged, which allows us to draw some clear conclusions: (1) the diffusivity increases exponentially when the molecular size of the diffusing molecule decreases, (2) an increase of temperature, leading to more thermal agitation, increases the diffusivity, and (3) any factor that could lead to an increase of free volume (e.g., glass transition phenomenon) will contribute to an increase of diffusivity.

In order to validate this theory, recently, Kilburn et al. (2004) explored the applicability of positron annihilation lifetime spectroscopy (PALS) to probing the structure of amorphous carbohydrate water systems. PALS is a powerful technique by which the size of voids in dense materials can be determined, and the technique is currently finding increased application in the analysis of the void structure of polymer matrices (Wang et al. 2004). From the PALS analysis of a glassy carbohydrate matrix, Kilburn et al. observed that the average volume of hole of the carbohydrate matrix increases strongly with increasing water content, but unfortunately they did not evaluate water vapor diffusivity. This technique appears therefore very promising for assessing the free volume theory in amorphous food and nonfood materials.

This free volume theory could be used for food materials behaving like polymers, such as carbohydrate matrices, but is not well adapted to lipidic materials such as edible coating, where the noncrystalline fraction is liquid and not an amorphous polymer (Morillon et al. 2002). Indeed, the noncrystalline fraction of a material would differ between polymers and lipids, consisting, for the former, of amorphous fraction and of liquid fraction with regard to lipids.

Nano- to Microscopic Scales

Addition of impermeable (nano- or micro-) particles (e.g., crystalline particles) in dense matrix also reduces the effective diffusivity of matter. For example, in dense lipids-based edible coatings, addition of crystalline materials (25 % of icing sugar) dramatically decreased the water vapor permeability coefficients of the coating. In these barriers, icing sugar addition induced structural reinforcement and limited the number of defects (that produces some preferential pathways for diffusion) and leads to a 50-fold decrease in the moisture diffusivity (Bourlieu 2007). The effect of the addition of impermeable particles (that do not take part in the diffusion process) in edible coatings on their diffusivity could be predicted through simple mathematical relationship as follows (Falla et al. 1996):

$$\frac{D_{ss}}{D_0} = \frac{2(1 - \phi)}{2 + \phi} \tag{4}$$

where D_0 is the diffusion coefficient through the continuous media, D_{ss} is the diffusion coefficient through the composite media, and ϕ is the volume fraction of the particles in the composite material.

Recently, Bilbao-Sainz and co-authors (Bilbao-Sainz et al. 2011a, b) reinforced HPMC film by addition of cellulose whiskers (HPMC/filler ratios in the dried films of 3:0.08 and 3:0.4). The composite films presented reduced WVP due to a lower affinity for water, whereas the water D_{eff} values were not affected by the filler. The impact of nanowhiskers is, in this case, due only to a reduced water sorption and not to an increase of tortuosity (i.e., physical barrier effect), contradicting what is usually stated and predicted by mathematical models such as that of Eq. 4.

When these impermeable particles are nanoparticles, their geometry and, especially, their aspect ratio is almost as important as the volume fraction of the particles. Indeed, for ribbon-like nanoparticles, even at low volume fraction, if a complete delamination and dispersion of the pellets within the matrix (exfoliation) is obtained, a tortuosity effect may be at the origin of the diffusion decrease. As far as the particles act as impermeable obstacles, the presence of filler introduces a longer diffusive path for the penetrant. Several mathematical models have been proposed to predict this effect. The model of Nielson (1967) was proved to well describe the reduction of the relative permeability by taking account of the volume fraction and the aspect ratio of particles and considering that particles are oriented parallel to the surface of the film (Choudalakis and Gotsis 2009). This model was further improved by Bharadwaj to take into account the possible random orientation of nanoparticles (Bharadwaj 2001):

$$\frac{D_{composite}}{D_{matrix}} = 1 \Big/ 1 + \frac{\bar{\alpha}}{2}\varphi\left(\frac{2}{3}\right)\left(S + \frac{1}{2}\right) \tag{5}$$

where $\bar{\alpha}$ and φ are, respectively, the number average aspect ratio of particles and the volume fraction of the particles. S is an order parameter defined as

$$S = \frac{1}{2}\langle 3\cos^2\theta - 1 \rangle \tag{6}$$

where θ represents the angle between the direction of preferred orientation and the surface of the film. The orientation parameter ranges from a value of -0.5 for a system, where the long axis of the filler is oriented parallel to the permeation direction (no tortuosity), to a value of 1, where it is oriented perpendicular to the permeation direction (maximum tortuosity), with a value of 0 representing random orientation.

This model was largely applied for synthetic polymers (Choudalakis and Gotsis 2009) and was successfully applied recently to wheat gluten films (Angellier-Coussy et al. 2012).

Micro- to Macroscopic Scales

Porosity

In a porous product, we usually consider that liquid diffusion occurs in the continuous, solid matrix and that vapor diffusion occurs in the gas phase, corresponding to gas bubbles of various sizes and shapes embedded in the solid matrix and sometimes connected (open pore). Since liquid diffusivity is 10^{-4} orders of magnitude lower than water vapor diffusivity (Gekas 1992), the overall effective moisture diffusivity may vary in large orders of magnitude depending on the ratio of open pores to solid matter—that is, depending on porosity value (Fig. 1). Decreasing the porosity of the cereal compartment (from 86 to 52 %) in a composite food reduced significantly the internal moisture diffusivity by at least a factor 2 (Roca et al. 2007). However, as far as we know, the relationship between moisture diffusivity and porosity change during remoistening or dehydration of the product was never modeled.

The main bottleneck to overcome in establishing the relationship between moisture diffusivity and porosity is the in situ measurement of this porosity. Indeed, most of the techniques required the drying of the sample (e.g., pycnometry, scanning

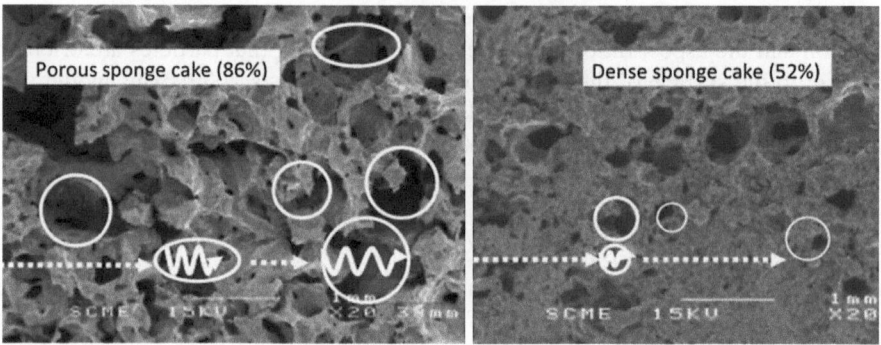

Fig. 1 Schematic representation of moisture transport mechanisms in porous cereal-based products as a function of percentage of porosity (*continuous lines* water vapor diffusion; *dotted lines* liquid water diffusion; *circles* pore walls) (adapted from Roca et al. 2008)

electronic microscopy, etc.) and, consequently, the evolution of porosity as a function of product moisture content could not be established. However, this relationship is essential as pointed out by the work of Guillard et al. (2003c) on sponge cake observed at various RH by using Environmental SEM (Fig. 8). Recently, new techniques, such as X-rays micro- or nanotomography or confocal microscopy, have emerged, coming from the material science field, and have been adapted in food science to the in situ investigation of porosity in real conditions (i.e., upon hydration kinetics, for example). For instance, Lim and Barigou (2004) have performed imaging, visualization, and analysis of the three-dimensional (3-D) cellular microstructure of a number of food products (aerated chocolate, mousse, marshmallow, and muffin) using X-ray microcomputed tomography (XMT). van Dalen et al. (2009) have performed such an analysis using Synchrotron equipment on cereal-based products. The obtained results demonstrate that Synchrotron microtomography can be used to reveal the internal microstructure of the lamellae separating the gas cells and to analyze the internal porosity and the size and shape of the starch granules in porous biscuits and crackers. Moreno–Atanasio et al. (2010) presented in their review a summary of the major applications in which computer simulations are explicitly coupled with XMT in the area of granular and porous materials.

Recently, in-depth investigations of the relationship between food structure and moisture transfer in cellular food solids were attempted. Esveld et al. (2012a) developed a pore scale network model using finite elements modeling to predict the dynamics of moisture diffusion into complex foods like bread, crackers, and cereals. The morphological characteristics of the sample, including the characteristics of each cellular void and the open pore connections between them, have been determined from X-ray microtomography data by means of 3-D image analysis techniques. In the study of Esveld et al. (2012a), the 3-D network allows simulation of the water vapor transport properties in the pores of the product without need of numerous effective parameters. As far as we know, this recent study brings significant advances in the food science field in the modeling of the relationship between food structure and mass transfer properties.

However, automatic quantitative 3-D image analyses are sometimes difficult to obtain due to the phase-contrast of the images (van Dalen et al. 2009). XMT is thus exclusively devoted to the structural investigation of porous products where the difference in phase density (gas vs. solid) is large enough. Even if promising, as far as we know, this technique has not yet been applied for the in situ evaluation of porosity change in porous food products upon moisture migration.

Similarly, the evolution of bench top NMR characteristics has resulted in the use of NMR relaxation times to build up contrast images (MRI) and their application to the investigation of food structure (Hills 1998; Mariette 2009). However, MRI images are affected both by change in molecular structure and by water content, which makes interpretation of T1 and T2 relaxation-time-based images more difficult. Despite this limit, applications of MRI protocols have led to the description of food matrices at the microscopic scale (100 μm up to several millimeters) during processing operations or storage periods inducing moisture changes (Ramos-Cabrer et al. 2006; Weglarz et al. 2008).

Elastic Properties of Food

The use of novel enabling technology that yields unprecedented insights into the physical transport phenomena that occur during food processing may lead to unexpected results. In the last 10 years, the use of MRI has permitted putting in evidence several cases of moisture migration that apparently violate Fick's law. Indeed, traditionally, in food science, diffusion is described by Fick's law, in which the mass flux is linear with the gradient in moisture content. A few papers have reported deviations of moisture transport from Fick's law during the drying of food products. In these papers, the moisture content in the product center increased during drying, whereas the average moisture content decreased with time, proving that water goes out of the product. This phenomenon was reported by Johnson et al. (1998) in plaintain, Courtois et al. (2001) in rice, Stapley et al. (1997) in steamed wheat grains, and recently, by Jin et al. (2012) in broccoli. Courtois et al. and Stapley et al. considered the internal structure of products as a reason for deviation and cooking under pressure. Jin et al. went further by hypothesizing and proving that differences in elasticity properties between the core and the skin of the product lead to gradients in elastic stress that provoke moisture migration against the gradient. In other words, shrinkage and deformation of the skin cause an internal pressure gradient, which results in a temporarily pressure-driven moisture transport toward the center of the product (van der Sman 2007; Okuzono and Doi 2008). During drying, the skin dries quickly, and due to the low moisture content, the skin is more elastic and causes a center-directed moisture transport (van der Sman 2007). Jin et al. put in evidence this phenomenon by pretreating their product before drying (e.g., peeling, blanching, freezing) to change the elasticity properties of the core, as compared with the peripheral tissues. This result means that drying models must be extended, with the addition of a stress-driven diffusion term included as proposed by Okozuno and Doi (Table 2).

These papers underline the importance of taking into account food structure heterogeneity to make accurate drying kinetics predictions. In addition, they proved that it would be probably more proper to relate the moisture transport to gradients in chemical potential (or equivalently, water activity or swelling pressure) instead of moisture content. If formulated in terms of the proper thermodynamic potential, there would exist no anomaly and deviation from Fick's law as evidenced in the studies mentioned above.

Multilayer Foods

Simulating moisture transfer in macroscopic structures such as multidomain systems is very useful for predicting in advance the product shelf life. Use of water sorption and diffusivity properties in such a model permit simulation of average moisture content as a function of time and, also, local distribution profile as a function of distance from interface (Fig. 2). However, the accuracy of the moisture transfer prediction is very dependent on the food structure. Multidomain foods are characterized by levels of structure of different scales: first, the internal structure of each compartment involved in the system (e.g., porous cereal-based product associated with a moist, dense wet filling) and, second, a macroscopic structure characterizing the association of the different compartments. Illustrating this last item, one may easily imagine that the contacts between compartments in a mixed salad or in a pizza or fresh pastry filled with cream are completely different and can strongly affect the moisture transfer at interface. In order to investigate the impact of the interface on water transfer kinetics, different systems representing the diversity of composite food products were studied by Roca et al. (2008a, b, c):

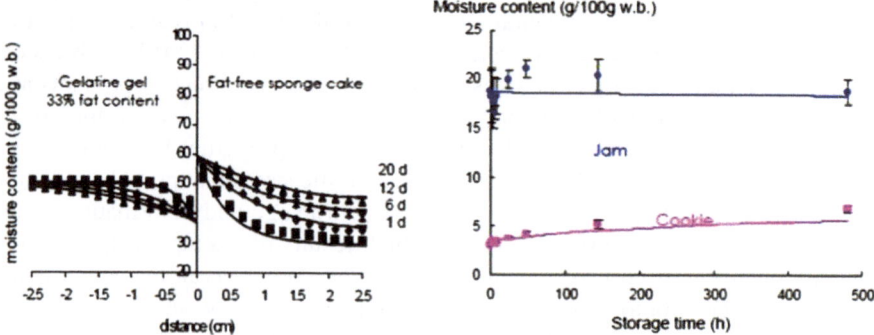

Fig. 2 Example of experimental and predicted data obtained when modelling moisture transfer in multi-domain structure

- two compartments in intimate contact (called *perfect contact*). In this case, the surface of the two compartments in contact is supposed to be instantaneously in water activity equilibrium;
- two compartments separated by a barrier film (the contact is supposed to be perfect between each compartment);
- two compartments separated by a thin gas layer (simulating the case of a mixed salad, where compartments are not completely stuck to each other). The thickness of the film layer may vary from several centimeters and millimeters and may tend to zero to mimic the system in perfect contact.

The nature of the interface (ideal or nonideal contact) between compartments strongly affected the water migration predictions. When the remoistening of a sponge cake in contact with an agar gel was simulated, the predicted moisture contents during storage were significantly lower in the case of imperfect contact (air gap), as compared with predictions for a system in direct contact (Roca et al. 2008a, b, c), even for very thin air layer thickness (around 1 mm). The shelf life of the product consequently increases from 2.8 to 3.2 days when a 1-mm air layer is present at the interface in the composite food. During simulation of temperature abuse during storage (1 day at 5 °C, then 0.5 day at 20 °C, and finally, 5.5 days at 20 °C) of their sponge cake/agar gel system, the same authors noticed that the presence of an air gap between the two compartments of the multidomain system reduced the negative effect of temperature fluctuations. The difference of sensibility to temperature depending on the type of contact may be related to the differential impact of temperature on water molecular diffusion (in the solid phases in perfect contact) and on water vapor diffusion occurring in the air gap (nonperfect contact). The air gap would behave like an "insulating" layer for moisture transfer.

It is thus crucial to consider the assembly of compartments in multidomain food systems, due to the importance of the macrostructure of the system for moisture transfer.

Presence of a Dense Structure at the Food Surface (Case of the Crust on Cereal-Based Product or of a Barrier Layer)

As has previously been discussed, reducing food porosity tends to slow down water diffusion by decreasing the value of effective diffusivity in the product. This result lets us think that the presence of a crust (such as, e.g., a bread crust or cake crust) on the top of a porous cereal-based structure might reduce the moisture transfer from the surrounding atmosphere or from the linked moist compartment into the core of the product. Armed with this hypothesis, Roca et al. (2006) simulated the case of a crust of a different thickness on a sponge cake associated with a wet compartment, agar gel. The system was considered as three compartments: sponge cake/crust/agar gel. The crust was considered to have the same

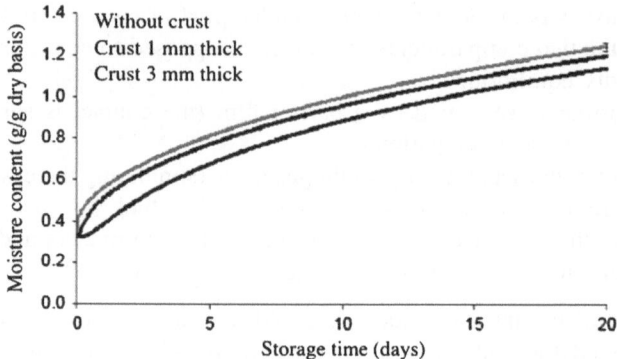

Fig. 3 Influence of presence or absence of crust on the cereal-based compartment involved in a multi-domain food where the cereal-based compartment (sponge cake of initial aw of 0.84) is in contact with a moist filling (agar gel of initial aw of 0.99)

characteristics of diffusion as the densest sponge cake formulation studied (i.e., 52 %, instead of 86 % for the soft interior). Simulations show that the presence of this crust, even when thin, significantly decreases the water quantity transferred from the gel to the sponge cake (Fig. 3).

The spatial structural heterogeneity of the food structure may thus be considered in the prediction of moisture transfer in food.

Multidomain Food with Edible Film

As was explained above, edible lipid-based films are an efficient means for limiting the moisture transfer between two components of contrasted a_w. A modeling approach permits prediction of the composite food product shelf life for a given structural integrity of the barrier film, optimizing composite food product configuration (thickness and chemical composition of the barrier film, initial a_w of the components,...), and limiting the number of experimental moisture migration studies. Mathematical models can simulate, for example, the variations in the period of acceptability of the product with increasing barrier film thicknesses, as described in Bourlieu et al. (2006).

Integration of the Various Scales: The Example of Hydrophobic Dense Edible Coatings

In the previous section, we presented the studies that have tentatively described and formalized (i.e., modeled) the relationship between structural and mass transfer properties in food. We pointed out that, in food science, the relationship between structure and transfer is rarely quantified and modeled. This illustrated well the complexity of the food materials and the difficulty of investigating in situ and in real conditions of use or storage the evolution of food structure. The scarce relationships established between structural and mass transfer properties were done at only one scale of observation, without any integration of the multiscale structure of the product. However, this multiscale structure is responsible for the main mechanisms governing mass transfer. For example, the structure at a molecular or macromolecular scale governs the local diffusion of gas or vapor, which is also affected by the nanoscale (presence of impermeable particles that increase the tortuosity of the matrice) and by the macrostructure of the product (multilayered structures). Most of the time, diffusion is characterized at a macroscopic scale by measuring an overall apparent diffusion coefficient. The apparent diffusion coefficient is a macroscopic property and cannot describe the different mechanisms of migration that prevail in food. A multiscale analysis of the relationship between the structure and mass transfer properties is thus required. Some attempts at multiscale analysis were carried out on lipid-based edible coatings used as moisture barriers, probably because of the relatively simpler structure of these dense materials, as compared with biological products or other foodstuffs. This section is devoted to the description of these attempts.

In addition to intrinsic chemical composition (presence of polar components, hydrocarbon chain length, number of unsaturations), the physical state (crystalline polymorphic organization) and microstructure (network of crystals, orientations of these crystals) of edible hydrophobic films influence its macroscopic water barrier properties and, more specifically, the water vapor permeability. Several authors have thus proposed a multiscale characterization of these types of edible films and have tried to relate the resulting macroscopic water barrier properties to the various levels of organization of the matrix (Martini et al. 2006) (see Fig. 1).

Bourlieu et al. (2009a) synthetized technical tristearin (TC18) with increasing acetylation degree (TC18 = 0 %, A2 = 47 %, A1 = 66 % acetyl mol·mol^{-1} esterified chains). The moisture sorption isotherms reported for the three matrices

V. Guillard et al., *Food Structure and Moisture Transfer*, SpringerBriefs in Food, Health, and Nutrition, DOI: 10.1007/978-1-4614-6342-9_4, © The Author(s) 2013

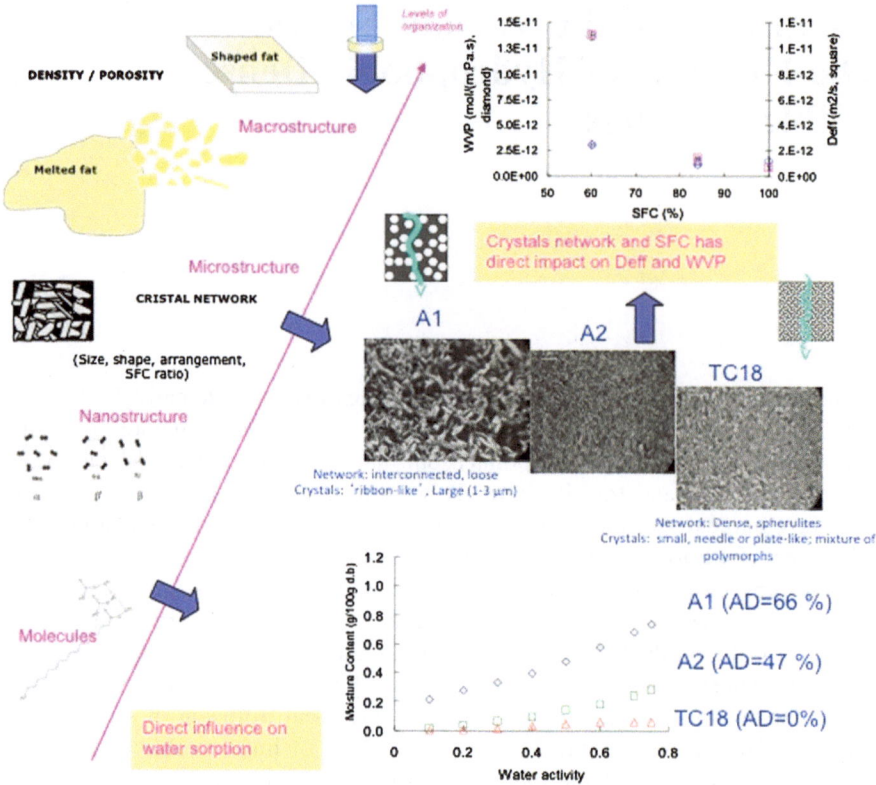

Fig. 1 Levels of structural organization of dense hydrophobic films and their influence on global moisture transfer (results of WVP and water vapor sorption taken and figures adapted from Bourlieu et al. 2010)

were directly influenced by this chemical composition and increased with acetylation degree, which induced higher polarity and liquid content. This acetylation degree also influenced the nanostructure and solid fat content (SFC) of the matrices: Polarized light micrographs displayed dense crystal networks for TC18 and A2, with some spherulitic organization of small-size, needle-like, or plate-like crystals for TC18 and anisotropic organization of small-size crystals for A2; A1 presented a looser interconnected network of large "ribbon-like" structures and much lower SFC (60 % vs. 77 % and 100 % for A2 and TC18, respectively). These differences in nano- and microstructures directly impacted the macroscopic property, D_{eff}, which values increased from 0.07 to 0.14 and 1.1 × 10^{-11} m^2 s^1 for TC18, A2, and A1. The D_{eff} values increased exponentially with the acetylation degree.

The resulting water vapor permeability, a macroscopic property, was also directly influenced by the molecular and nanostructure in the aforementioned study. Indeed, WVP was found to increase with acetylation degree (from A1, with an acetylation degree of 47 %, to A2, with 66 %). But for this parameter, the structure at the micro- and macroscale was also peculiarly important. The apparition of structural defects in TC18 did not lead to a decrease in WVP, since it was expected due to its very high SFC. These structural defects were revealed by SEM cross-sections in TC18 and were induced by the very bad mechanical properties (due to high SFC) of this material, which favored the creation of cracks during WVP measurement. These authors, however, did not propose any mechanistic model linking moisture sorption, moisture effective diffusivity, and the various levels of organization (molecular, nanoscopic, microscopic) investigated.

Most recent works on hydrophobic edible films present the same drawback, with good characterization of the various levels of organization of the edible film structure using laser light scattering granulometry, SEM cross-sections, AFM, X-ray, surface hydrophobicity measurements, and as output of the WVP of the edible film. However, no suggestion of model linking these microstructural levels to the macroscopic diffusion or sorption of water in the matrix is being made (Jimenez et al. 2010, 2012; Kokoszka et al. 2010).

Concluding Remarks

The objective of this chapter was to give an overview of the different studies conducted in food science whose aim was to describe and model the relationship between food structure and mass transfer properties. Although the impact of food structure on moisture transfer is now well established, the link between some structural properties and the transport coefficient has scarcely been quantified and formalized. The most advanced works on the topic have been done for dense edible films and coatings, because methodologies and modeling tools developed in material science and engineering can be more easily translated to these materials than to complex food materials. When the relationship and modeling of structure/ transfer properties have been approached, most of the time, it has been at the molecular or nanoscopic scale. Multiscale analysis is lacking, probably because of the gap existing between scales of structural observation (from the molecular, via the nano- to the micro- and macroscale) and the scale used for characterizing and modeling the mass transfer properties (only the macroscopic scale, if we except the peculiar case of self-diffusion, which is out of the scope of the topic discussed here). Although new tools, such as X-ray microtomography, now permit the characterization of food structure with high resolution, there is still a lack of methodologies for measuring a local diffusion at a macromolecular and nanoscopic scale with local gradients of concentrations.

References

Adamson AW (1967) Physical chemistry of surfaces. Interscience, New York

Adedeji AA, Ngadi MO (2009) 3-D imaging of deep-fat fried chicken nuggets breading coating using X-ray micro-CT. Int J Food Eng 5(4). http://works.bepress.com/akinbode_adedeji1/1

Anderson RB (1946) Modifications of the BET equation. J Am Chem Soc 68:689–691

Andrieu J, Jallut C, Stomatopoulos AA, Zafiropoulos M (1988) Identification of water apparent diffusivities for drying of corn based extruded pasta Int. In: Drying symposium, Versailles, France

Andrieu J, Stamatopoulos AA (1986) Durum-wheat pasta drying kinetics. Lebensm Wiss Technol 19(6):448–456

Angellier-Coussy H, Gastaldi E, Correa F, Gontard N, Guillard V (2012) Nanoparticle size and water diffusivity in nanocomposite agro-polymer based films. Eur Polym J (in press), http://dx.doi.org/10.1016/j.eurpolymj.2012.11.006

Angellier-Coussy H, Gastaldi E, Gontard N, Guillard V (2011) Influence of processing temperature on the water vapour transport properties of wheat gluten based agromaterials. Ind Crops and Products 33:457–461

ASTM E-96 (1980) Standard test methods for water vapor transmission of materials. ASTM Book of Standards

Baik OO, Marcotte M (2002) Modelling the moisture diffusivity in a baking cake. J Food Eng 56:27–36

Barbosa-Canovas G, Vega-Mercado H (1996) Dehydration of foods. In: Barbosa Canovas G, Vega-Mercado H (eds) Chapman and Hall, New York, 329 pp

Batista LM, da Rosa CA, Pinto LAA (2007) Diffusive model with variable effective diffusivity considering shrinkage in thin layer drying of chitosan. J Food Eng 81(1):127–132

Baucour P, Daudin JD (2000) Development of a new method for fast measurement of water sorption isotherms in the high humidity range—validation on gelatine gel. J Food Eng 44(2):97–107

Bell LN, Labuza TP (2000) Moisture sorption isotherms. In: Bell LN, Labuza TP (eds) Moisture sorption. Practical aspects of isotherm measurement and use. Paperback, Saint Paul, pp 70–98

Berger D, Pei DCT (1973) Drying of hygroscopic capillary porous solids. A theoretical approach. Int J Heat Mass Transf 1:293–302

Bertuzzi MA, Castro Vidaurre EF, Armada M, Gottifredi JC (2007) Water vapor permeability of edible starch based films. J Food Eng 80(3):972–978

Bharadwaj RK (2001) Modeling the barrier properties of polymer-layered silicate nanocomposites. Macromolecules 34(26):9189–9192

Bilbao-Sainz C, Bras J, Williams T, Sénechal T, Orts W (2011a) HPMC reinforced with different cellulose nano-particles. Carb Polym 86:1549–1557

Bilbao-Sainz C, Wood R, William T, McHugh TH (2011b) Composite Edible Films Based on Hydroxypropyl Methylcellulose Reinforced with Microcrystalline Cellulose Nanoparticles. J Agric Food Chem 58(6):3753–3760

Biquet B, Guilbert S (1986) Relative diffusivities of water in model intermediate moisture foods. Lebensm-Wiss Technol 19(3):208–214

Biquet B, Labuza TP (1988) Evaluation of the moisture permeability characteristics of chocolate films as an edible moisture barrier. J Food Sci 53:989–998

Bonazzi C, Ripoche A, Michon C (1997) Moisture diffusion in gelatin slabs by modeling drying kinetics. Dry Technol 15(6-8):2045–2059

Boudhrioua N, Bonazzi C, Daudin JD (2003) Estimation of moisture diffusivity in gelatin-starch gels using time-dependent concentration-distance curves at constant temperature. Food Chem 82:139–149

Bourlieu C, Guillard V, Powell H, Vallès-Pàmies B, Guilbert S, Gontard N (2006) Performance of lipid-based moisture barriers in intermediate water activity food products. Eur J Lipid Sci Tech 108:1007–1020

Bourlieu C (2007) Edible moisture barriers: characterization, modelling of potential and limits in composite food products stabilization. Food Engineering - Food Science. University of Montpellier, Montpellier, phD, 309 pp

Bourlieu C, Guillard V, Powell H, Vallès-Pàmies B, Guilbert S, Gontard N (2008) Modelling and control of moisture transfers in high, intermediate and low aw composite food. Food Chem 106:1350–1358

Bourlieu C, Ferreira M, Barea B, Guillard V, Villeneuve P, Guilbert S, Gontard N (2009a) Moisture barrier and physical properties of acetylated derivatives with increasing acetylation degree. Eur J Lipid Sci Tech 111(5):489–498

Bourlieu C, Guillard V, Vallès-Pamiès B, Gontard N (2009b) Edible moisture barriers: Uses promises and limits in food products shelf-life extension. Crit Rev Food Sci 49(5):474–499

Bourlieu C, Guillard V, Ferreira M, Powell H, Valles-Pamies B, Guilbert S, Gontard N (2010) Effect of cooling rate on the structural and moisture barrier properties of high and low melting point fats. J Am Oil Chem Society 87:133–145

Bove L, D'Aniello C, Gorrasi G, Guadagno L, Vittoria V (1996) Transport properties of dichloromethane in glassy polymers. 6. Poly(ethylene Terephthalate). J Appl Polym Sci 62:1035–1041

Bruce DM (1985) Exposed-layer barley drying, three models fitted to new data up to 150°C. J Agric Eng Res 32:547–554

Brunauer S, Deming LS, Teller E (1940) A theory of the van der Waals adsorption of gases. J Am Oil Chem Soc 62(7):1723–1732

Brunauer S, Emmett PH, Teller E (1938) Adsorption of gases in multi molecular layers. J Am Oil Chem Soc 62:1723–1732

Burden RL, Faires JD (1997) Numerical analysis, 6th edn. Brooks/Cole Publishing Company, Florence, pp 682–698

Buvanasundaram K, Mukai N, Tsukada T, Hozawa M (1994) Experimental and simulation study on drying of food gel. In Mujumdar, A.S.: Drying 94. New York, NY, Hemisphere, vol. 2:1291–1298

Chakraverty A (1984) Thin-layer characteristics of cashew nuts and cashew kernels. In: Mulumdar AS (ed) Drying'84. Hemisphere, Washington, DC, pp 396–400

Chao RR, Rizvi SSH (1988) Oxygen and water-vapor transport through polymeric film—a review of modeling approaches. Acs Symp Ser 365:217–242

Chen PL, Long R, Ruan R, Labuza TP (1997) Nuclear magnetic resonance studies of water mobility in bread during storage. Lebensm Wiss Technol 30:178–183

Chiang WC, Petersen JN (1987) Experimental measurement of temperature and moisture profiles during apple drying. Dry Technol 5(1):25–49

Chirife J, Boquet R, Ferro Fontan C, Iglesias HA (1983) A new model for describing the water sorption isotherm of foods. J Food Sci 48(4):1382–1383

Chirife J, Iglesias HA (1978) Equations for fitting water sorption isotherms of foods. 1. Rev J Food Technol 13(3):159–174

Chivrac F, Angellier-Coussy H, Guillard V, Pollet E, Avèrous L (2010) How does water diffuse in starch/montmorillonite nano-biocomposite materials? Carb Polym 82(1):128–135

Choudalakis G, Gotsis AD (2009) Permeability of polymer/clay nanocomposites: a review. Eur Polymer J 45(4):967–984

Chu ST, Hustrulid A (1968) General characteristics of variable diffusivity process and the dynamic equilibrium moisture content. T Asae 11:709–715

Colon G, Aviles EI (1993) Transport mechanisms in the drying of granular solids. In: Mujumdar AS (ed) Food dehydration, vol 2. Hemisphere Publishing, New York

Courtois F, Archila MA, Bonazzi C, Meot JM, Trystram G (2001) Modeling and control of a mixed-flow rice dryer with emphasis on breakage quality. J Food Eng 49(4):303–309

Crank J (1975) The mathematics of diffusion, 2nd edn. Clarendon Press, Oxford, 414 pp

Dantas LB, Orlande HRB, Cotta RM (2003) An inverse problem of parameter estimation for heat and mass transfer in capillary porous media. Int J Heat Mass Transf 46(9):1587–1598

Daudin JD (1983) Calcul des cinétiques de séchage par l'air chaud des produits biologiques solides. Sciences des Aliments 3:1–36

De Boer JH (1953) The dynamical character of adsorption. Clarendon Press, Oxford

Del Valle FR, Nickerson JTR (1968) Salting and drying of fish. 3. Diffusion of water. J Food Sci 33:499–503

Dhoot SN (2004) Sorption and transport of gases and organic vapors in poly(ethylene terephtalate). PhD thesis, University of Texas at Austin

Djelveh G, Gros JB, Bories B (1989) An improvement of the cell diffusion method for the rapid determination of diffusion constants in gels or foods. J Food Sci 54(1):166–169

Djelveh G, Petit M, Gros JB (1988) Influence of sodium chloride concentration potassium, nitrate end temperature on the apparent diffusion coefficient of chloride ions through agar gels. Lebensm Wiss Technol 21:103–107

Doulia D, Tzia K, Gekas V (2000) A knowledge base for the apparent mass diffusion coefficient (Deff) of foods. Int J Food Prop 3(1):1–14

Duda J, Zielinsky J (1996) Free volume theory in diffusion in polymers by: Neogi, P. CRC, London, pp 143–173

Esveld DC, van der Sman RGM, van Dalen G, van Duynhoven JPM, Meinders MBJ (2012a) Effect of morphology on water sorption in cellular solid foods. Part I: pore scale network model. J Food Eng 109(2):301–310

Esveld DC, van der Sman RGM, Witek MM, Windt CW, van As H, van Duynhoven JPM, Meinders MBJ (2012b) Effect of morphology on water sorption in cellular solid foods. Part II: sorption in cereal crackers. J Food Eng 109(2):311–320

Falla WR, Mulski M, Cussler EL (1996) Estimating diffusion through flake-filled membranes. J Membr Sci 119(1):129–138

Ferro-Fontan C, Chirife J, Sancho E, Iglesias HA (1982) Analysis of a model for water sorption phenomena in foods. J Food Sci 47:1590–1594

Fick A (1855) On liquid diffusion. London, Edinburgh and Dublin Philos Mag J Sci 10:30–39. (Abstracted by the author from the German original : Über Diffusion, Poggendorff's Annalen der Physik und Chemie, (1855), 94:59–86)

Floros JD, Chinnan MS (1989) Determining the diffusivity of sodium hydroxide through tomato and capsicum skins. J Food Eng 9:129–141

Fortes M, Okos M (1981a) A non-equilibrium thermodynamics approach to transport phenomena in capillary porous media. T ASAE 24(3):756–760

Fortes M, Okos MR (1981b) Nonequilibrium thermodynamics approach to heat and mass-transfer in corn kernels. T ASAE 24(3):761–769

Gardner TQ, Falconer JL, Noble RD (2002) Adsorption and diffusion properties of zeolite membranes by transient permeation. Desalination 149(1-3):435–440

Gekas V (1992) Transport phenomena of foods and biological materials, 1st edn. CRC Press, London, 237 pp

Gencturk B, Bakshi AS, Hong YC, Labuza TP (1986) Moisture transfer properties of wild rice. J Food Process Eng 8:243–253

Ghosh V (2003) Moisture Migration through chocolate—flavoured coatings. Ph.D. thesis, Pennsylvania State University, USA

Gill EP, Murra W, Wright MH (1981) Practical optimization. Academic Press, Inc, New York 401 pp

Gontard N, Guilbert S, Cuq JL (1992) Edible wheat gluten films—influence of the main process and environmental-conditions on thermal, mechanical and barrier properties. In: Abstracts of papers of the American Chemical Society. pp 204, 217-AGFD

Gouanve F, Marais S, Bessadok A, Langevin D, Metayer M (2007) Kinetics of water sorption in flax and PET fibers. Eur Polym J 43(2):586–598

Guggenheim EA (1966) Applications of statistical mechanics. Clarendon Press, Oxford

Guillard V, Broyart B, Bonazzi C, Guilbert S, Gontard N (2003a) Evolution of moisture distribution during storage in a composite food modelling and simulation. J Food Sci 68(3):958–966

Guillard V, Broyart B, Bonazzi C, Guilbert S, Gontard N, (2003b) Modelling of moisture transfer in a composite food: dynamic water properties in an intermediate a(w) porous product in contact with high aw filling. Chem Eng Res Des 81(A9):1090–1098

Guillard V, Broyart B, Bonazzi C, Guilbert S, Gontard N (2003c) Moisture diffusivity in sponge cake as related to porous structure evaluation and moisture content. J Food Sci 68(2):555–562

Guillard V, Broyart B, Bonazzi C, Guilbert S, Gontard N (2003d) Preventing moisture transfer in a composite food using edible films: experimental and mathematical study. J Food Sci 68(7):2267–2277

Guillard V, Bonazzi C, Guilbert S, Gontard N (2004a) Edible acetylated monoglyceride films: Effect of film-forming technique on moisture barrier properties. J Am Oil Chem Soc 81:1053–1058

Guillard V, Broyart B, Bonazzi C, Guilbert S, Gontard N (2004b) Effect of temperature on moisture barrier efficiency of monoglyceride edible films in cereal-based composite foods. Cereal Chem 81(6):767–771

Guillard V, Broyart B, Bonazzi C, Guilbert S, Gontard N (2004c) Moisture diffusivity and transfer modelling in dry biscuit. J Food Eng 64(1):81–87

Gürtas FS (1994) Low temperature drying of cultured mushroom (A. biporus). M. Sc. thesis, Istanbul Technical University, Istanbul

Halsey G (1948) Physical adsorption on non-uniform surfaces. J Chem Phys 16:931–937

Hebbar HU, Rastogi NK (2001) Mass transfer during infrared drying of cashew kernel. J Food Eng 47:1–5

Henderson SM (1974) Progress in developing equation the thin-layer drying. Trans Am Soc Agric Eng 17:1167–1172

Henderson SM, Pabis S (1961) Grain drying theory I: temperature effect on drying coefficient. J Agric Res Eng 4:169–174

Higughi T, Aguiar A (1959) A study of permeability to water vapour of fats, waxes, and other coating materials. J Am Pharm Assoc 48(10):574–583

Hills B (1998) Magnetic resonance imaging in food science. A Wiley Interscience publication, New York

Hoch G, Chauhan A, Radke CJ (2003) Permeability and diffusivity for water transport through hydrogel membranes. J Membr Sci 214(2):199–209

Hsu KH (1983) A diffusion-model with a concentration-dependent diffusion-coefficient for describing water-movement in legumes during soaking. J Food Sci 48(2):618–622

Hussain A, Chen CS, Clayton JT (1973) Simultaneous heat and mass diffusion in biological materials. J Agric Eng Res 18:343–354

Hutchinson D, Otten L (1983) Thin-layer air drying of soybeans and white beans. J Food Technol 18:507–524

Hwang SS, Cheng YC, Chang C, Lur HS, Lin TT (2009) Magnetic resonance imaging and analyses of tempering processes in rice kernels. J Cereal Sci 50(1):36–42

Jimenez A, Fabra MJ, Talens P, Chiralt A (2010) Effect of lipid self-association on the microstructure and physical properties of hydroxypropyl-methylcellulose edible films containing fatty acids. Carb Polym 82:585–593

Jimenez A, Fabra MJ, Talens P, Chiralt A (2012) Effect of re-crystallization on tensile, optical and water vapour barrier properties of corn starch films containing fatty acids. Food Hydrocoll 26:302–310

Jin X, van Boxtel AJB, Gerkema E, Vergeldt FJ, van As HT, van Straten G, Boom RM, van der Sman RGM (2012) Anomalies in moisture transport during broccoli drying monitored by MRI? Faraday Discuss 158:1–11

Johnson PNT, Brennan JG, Addo-Yobo FY (1998) Air-drying characteristics of plantain (Musa AAB). J Food Eng 37(2):233–242

Karathanos VT, Kostaropoulos AE (1995) Diffusion and equilibrium of water in dough/raisin mixtures. J Food Eng 25:113–121

Karathanos VT, Reppa A, Kostaropoulos AE (1994) Air-drying kinetics of osmotically dehydrated fruits. In: Mujumdar, AS (ed) Drying'94, vol 1. Hemisphere, New York, pp. 871–878

Karathanos VT, Villalobos G, Saravacos GD (1990) Comparison of 2 methods of estimation of the effective moisture diffusivity from drying data. J Food Sci 55(1):218–223

Kerkhof Piet JAM (1994) The role of theoretical and mathematical modelling in scale-up. Dry Technol 12:1–46

Khalfaoui M, Knani S, Hachicha MA, Lamine AB (2003) New theoretical expressions for the five adsorption type isotherms classified by BET based on statistical physics treatment. J Colloid Int Sci 263:350–356

Kilburn D, Claude J, Mezzenga R, Dlubek G, Alam A, Ubbink J (2004) Water in glassy carbohydrates: opening it up at the nanolevel. J Phys Chem B 108(33):12436–12441

Kiranoudis CT, Maroulis ZB, Marinos Kouris V (1992) Model selection in air drying of foods. Dry Technol 10(4):1097–1106

Kiranoudis CT, Maroulis ZB, Marinos Kouris D (1995) Heat and mass transfer model building in drying with multiresponse data. Int J Heat Mass Transf 38(3):463–480

Koelsch CM, Labuza TP (1992) Functional, physical and morphological properties of methyl cellulose and fatty acid-based edible barriers. Lebensm-Wiss Technol 25:404–411

Kokoszka S, Debeaufort F, Lenarta A, Voilley A (2010) Liquid and vapour water transfer through whey protein/lipid emulsion films. J Sci Food Agric 90:1673–1680

Kulasiri D, Samarasinghe S (1996) Modelling heat and mass transfer in drying of biological materials: a simplified approach to materials with small dimensions. Ecol Model 86(2-3):163–167

Labuza TP, Acott K, Tatini SR, Lee RY, Flink J, McCall W (1976) Water activity determination—collaborative study of different methods. J Food Sci 41(4):910–917

Landmann W, Lovegren NV, Feuge RO (1960) Permeability of some fat products to moisture. J Am Oil Chem Soc 37:1–4

Lazarides HN, Gekas V, Mavroudis N (1997) Apparent mass diffusivities in fruit and vegetable tissues undergoing osmotic processing. J Food Eng 31:315–324

Leslie RB, Carillo PJ, Chung TY, Gilbert SG, Hayakawa K, Marousis S, Saravacos GD, Solberg M (1991) Water diffusivity in starch-based systems. In: Levine H, Slade L (eds) Water relationships in foods, advances in the 1980s and trends for the 1990s. Plenum Press, New York, pp. 365–390

Lim KS, Barigou M (2004) X-ray micro-computed tomography of cellular food products. Food Res Int 37(10):1001–1012

Litchfield JB, Okos M (1992) Moisture diffusivity in pasta during drying. J Food Eng 17:117–142

Lomauro CJ, Bakshi AS, Labuza TP (1985a) Evaluation of food moisture sorption isotherm equations. 1. fruit, vegetable and meat-products. Lebensm-Wiss Technol 18(2):111–117

Lomauro CJ, Bakshi AS, Labuza TP (1985b) Evaluation of food moisture sorption isotherm equations. 2. milk, coffee, tea, nuts, oilseeds, spices and starchy foods. Lebensm-Wiss Technol 18(2):118–124

Lovegreen NV, Feuge RO (1954) Permeability of acetostearin products to water vapour. J Agric Food Chem 2:558–563

Luikov AV (1966) Application of the irreversible thermodynamic methods to investigation of heat and mass transfer. Int J Heat Mass Transf 9:139–152

Luikov AV (1975) Systems of differential equations of heat and mass transfer in capillary-porous bodies (review). Int J Heat Mass Transf 18(1):1–4

Madamba PS, Driscoll RH, Buckle KA (1996) Thin-layer drying characteristics of garlic slices. J Food Eng 29:75–97

Mariette F (2009) Investigations of food colloids by NMR and MRI. Curr Opin Colloid Int Sci 14(3):203–211

Maroulis ZB, Tsami E, Marinos-Kouris D, Saravacos GD (1988) Application of the GAB model to sorption isotherm of fruits. J Food Eng 7:63–78

Marousis SN, Karathanos VT, Saravacos GD (1989) Effect of sugars on the water diffusivity in hydrated granular starches. J Food Sci 54(6):1496–1500

Marousis SN, Karathanos VT, Saravacos GD (1991) Effect of physical structure of starch materials on water diffusivity. J Food Process Preserv 15(3):183–195

Martini S, Kim DA, Ollivon M, Marangoni AG (2006) The water vapor permeability of polycrystalline fat barrier films. J Agric Food Chem 54:1880–1886

McHugh TH, Avenabustillos R, Krochta JM (1993) Hydrophilic edible films—modified procedure for water-vapor permeability and explanation of thickness effects. J Food Sci 58(4):899–903

McHugh TH, Krochta JM (1994) Permeability properties of edible films. In: Krochta JM, Baldwin EA, Nisperos-Carriedo MO (eds) Edible films and coatings to improve food quality. Technomic publishing company, Lancaster, pp 139–187

Moreno-Atanasio R, Williams RA, Jia X (2010) Combining X-ray microtomography with computer simulation for analysis of granular and porous materials. Particuology 8:81–99

Morillon V, Debeaufort F, Blond G, Capelle M, Voilley A (2002) Factors affecting the moisture permeability of lipid-based edible films: a review. Crit Rev Food Sci Nutr 42(1):67–89

Motarjemi Y (1988) A study of some physical properties of water in foodstuffs. Water activity, water binding and water diffusivity in minced meat products. Ph.D. thesis, Lund University, Lund, Sweden

Mourad M, Hemati M, Laguerie C (1994) Evolution of the technical quality of corn kernels submitted to various techniques of drying at different operating conditions. In: Mujumdar AS (ed) Drying 94, vol 1. Hemisphere, New York, pp 1023–1030

Mulet A (1994) Drying modelling and water diffusivity in carrots and potatoes. J Food Eng 22:329–348

Myers AW, Meyer JA, Rogers CE, Stannett V, Szwarc M (1961) Studies in the gas and vapor permeability of plastic films and coated papers. Part VI—the permeation of water vapour. TAPPI 44:58–64

Navarro-Tarazaga ML, Sothornvit R, Perez-Gago MB (2008) Effect of plasticizer type and amount on hydroxypropyl methylcellulose-beeswax edible film properties and postharvest quality of coated plums. J Agric Food Chem 56(20):9502–9509

Naylor T de V (1989) Permeation properties. In: Booth C, Price C (eds) Comprehensive polymer science, vol 2. Pergamon Press Inc., New York, pp 643–668

Neményi M, Czaba I, Kovács A, Jáni T (2000) Investigation of simultaneous heat and mass transfer within the maize kernels during drying. Comput Electron Agric 26(2):123–135

Nielsen LE (1967) Models for the permeability of filled polymer systems. J Macromol Sci-Phys 1:929–942

O'Callaghan JR, Menzies DJ, Bailey PH (1971) Digital simulation of agricultural drier performance. J Agric Eng Res 16:309–333

Okuzono T, Doi M (2008) Effects of elasticity on drying processes of polymer solutions. Phys Rev E 77(3):Part 1

Oliver L, Meinders MBJ (2011) Dynamic water vapour sorption in gluten and starch films. J Cereal Sci 54(3):409–416

Ollivon M, Adenier H (2003) Chapitre 8. Technologie du chocolat et produits. In: Graille J (ed) EdLipides et corps gras alimentaires. Editions Tec & Doc, Paris, pp 275–313

Oswin CR (1946) The kinetics of package life. III. The isotherm. J Chem Ind (Lond) 65:419–423

Ozdemir M, Devres YO (1999) The thin layer drying characteristics of hazelnuts during roasting. J Food Eng 42:225–233

Palou E, Lopez-Malo A, Argaiz A (1997) Effect of temperature on the moisture sorption isotherms of some cookies and corn snacks. J Food Eng 31:85–93

Park GS (1986) Transport principles: solution. Diffusion and permeation in polymer membranes. Reidel Publications, Holland

Park HJ, Chinnan MS (1995) Gas and water vapor barrier properties of edible films from protein and cellulosic materials. J Food Eng 25:497–507

Parti M, Dugmanics I (1990) Diffusion coefficient for corn drying. T ASAE 33:1652–1656

Peleg M (1993) Assessment of a semi-empirical four parameter general modem for sigmoid moisture sorption isotherms. J Food Process Eng 16:21–37

Philip JR, De Vries DA (1957) Moisture movement in porous materials under temperature gradients. Trans Amer Geophys Union 38:222–232

Quezada-Gallo JA (1999) Influence de la structure et de la composition de réseaux macromoléculaires sur les transferts de molécules volatiles (eau et arômes). Application aux emballages comestibles et plastiques. Ph.D. thesis, Université de Bourgogne, France

Raghavan GSV, Tulasidas TN, Sablani SS, Ramaswamy HS (1995) A method of determination of concentration dependent effective moisture diffusivity. Dry Technol 13(5-7):1477–1488

Ramos-Cabrer P, Van Duynhoven JPM, Timmer H, Nicolay K (2006) Monitoring of moisture regiodistribution in multicomponent food systems by use of magnetic resonance imaging. J Agric Food Chem 54:672–677

Renzetti S, Voogt JA, Oliver L, Meinders MBJ (2012) Water migration mechanisms in amorphous powder material and related agglomeration propensity. J Food Eng 110(2): 160–168

Rhim JW, Gennadios A, Weller CL, Hanna MA (2002) Sodium dodecyl sulfate treatment improves properties of cast films from soy protein isolate. Ind Crops Products 15:199–205

Roca E, Guillard V, Guilbert S, Gontard N (2006) Moisture migration in a cereal composite food at high water activity: effects of initial porosity and fat content. J Cereal Sci 43(2):144–151

Roca E, Broyart B, Guillard V, Guilbert S, Gontard N (2007) Controlling moisture transport in a cereal porous product by changing structural or formulation parameters. Food Res Int 40:461–469

Roca E, Adeline D, Guillard V, Guilbert S, Gontard N (2008a) Shelf-life and moisture transfer predictions in a composite food product: impact of preservation techniques Int J Food Eng 4(4):Art. 4

Roca E, Broyart B, Guillard V, Guilbert S, Gontard N (2008b) Predicting moisture transfer and shelf-life of multidomain food products. J Food Eng 86:74–83

Roca E, Guillard V, Broyart B, Guilbert S, Gontard N (2008c) Effective moisture diffusivity modelling versus food structure and hygroscopicity. Food Chem 106(4):1428–1437

Rogers CE (1965) Physics and chemistry of the organic solid state. In: Fox D, Labes MN, Weisser A (eds) Interscience, vol 2. New York, pp 509–635

Rogers CE (1985) Permeation of gases and vapours in polymers. In: Comyn J (ed) Polymer permeability. Elsevier Applied Science, London, pp 11–73

Roos YH (1995) Phase transitions in foods. Academic Press, San Diego 360 pp

Rougier T, Bonazzi C, Broyart B, Daudin JD (2007) Impact of lipid phase on water transfer in food. In: 3rd interamerican drying conference location, McGill University, Montreal, Canada, 21–23 Aug 2005. Dry Technol 25(1–3):341–348

Roussenova M, Murith M, Alam A, Ubbink J (2010) Plasticization, antiplasticization, and molecular packing in amorphous carbohydrate-glycerol matrices. Biomacromolecules 11:3237–3247

Ruiz Cabrera MA (1999) Détermination de la relation entre la diffusivité de l'eau et la teneur en eau dans les matériaux déformables à partir d'images RMN. Elaboration de la méthode avec des gels de gélatine et transposition à la viande. Ph.D. thesis, Université d'Auvergne, Clermont-Ferrand, 122 pp

Sabarez H, Price WE et al (1997) Modelling the kinetics of drying of d'Agen plums (Prunus domestica). Food Chem 60(3):371–382

Sabarez HT, Price WE (1999) A diffusion model for prune dehydration. J Food Eng 42(3):167–172

Sanni LO, Atere C, Kuye A (1997) Moisture sorption isotherms of Fufu and tapioca at different temperatures. J Food Eng 34:203–212

Schrader GW, Litchfield JB (1992) Moisture profiles in a model food gel during drying : measurement using magnetic resonance imaging and evaluation of the fickian model. Dry Technol 10(2):295–332

Serad GE, Freeman BD, Stewart ME, Hill AJ (2001) Gas and vapor sorption and diffusion in poly(ethylene terephthalate). Polymer 42:6929–6943

Sharaf-Eldeen YI, Blaisdell JL, Hamdy MY (1980) A model for ear corn drying. Trans Am Soc Agric Eng 23(1261-1265):1271

Shellhammer TH, Rumsey TR, Krochta JM (1997) Viscoelastic properties of edible lipids. J Food Eng 33:305–320

Shellhammer TH, Krochta JM (1997a) Chapter 17. Edible coatings and film barriers. In: Gunstone FD, Padley FB (eds) Lipid technologies and applications. Marcel Dekker, New-York, pp 453–479

Shellhammer TH, Krochta JM (1997b) Water vapor barrier and rheological properties of simulated and industrial milkfat fractions. T ASAE 40:1119–1127

Simatos D (2002) Propriétés de l'eau dans les produits alimentaires : activité de l'eau, digrammes de phases et d'états. In: Le Meste M, Simatos D, Lorient D (es) L'eau dans les aliments. Tec & Doc, Paris, pp 49–83

Singh RP, Lund DB, Buelow FH (1984) An experimental technique using regular regime theory to determine moisture diffusivity. In: MacKenna BM (ed) Engineering and food, vol 1. Elsevier Applied Science, London, pp 415–423

Smith SE (1947) The sorption of water vapour by high polymers. J Am Chem Soc 69(3):646–651

Stapley AGF, Hyde TM, Gladden LF, Fryer PJ (1997) NMR imaging of the wheat grain cooking process. Int J Food Sci Technol 32(5):355–375

Thompson TL, Peart RM, Foster GH (1968) Mathematical simulation of corn drying—a new model. Trans Am Soc Agric Eng 11:582–586

Thorvaldsson K, Janestad H (1999) A model for simultaneous heat, water and vapour diffusion. J Food Eng 40(3):167–172

Tien C (1994) Adsorption calculation and modeling. Butterworth-Heinemann, London

Timmermann EO, Chirife J (1991) The physical state of water sorbed at high activities in starch in terms of the GAB sorption equation. J Food Eng 13(3):171–179

Tong CH, Lund DB (1990) Effective moisture diffusivity in porous materials as a function of temperature and moisture-content. Biotechnol Prog 6(1):67–75

Tsujita Y (1992) The physical chemistry of membranes science and technology. In: Osaka Y, Nakaganta T (eds) Membrane science and technology. Dekker Marcel Inc, Basel

Tütüncü MA, Labuza TP (1996) Effect of geometry on the effective moisture transfer diffusion coefficient. J Food Eng 30:433–447

Ubbink J, Giardiello MI, Limbach HJ (2007) Sorption of Water by bidisperse mixtures of carbohydrates in glassy and rubbery states. Biomacromolecule 8(9):2862–2873

Ulku S, Uckan G (1986) Corn drying in fluidized beds. In: Mujumdar AS (ed) Drying 86, vol 2. Hemisphere, New York, pp 531–536

Vagenas GK, Karathanos VT (1993) Prediction of the effective moisture diffusivity in gelatinised food systems. J Food Eng 18:159–179

van Dalen G, Nootenboom P, Don A, den Adel R, Roijers E (2009) 3D imaging of the solid phase of porous bakery products using synchrotron X-ray microtomography. In Capasso V, Aletti G, Micheletti A (eds) Ecs 10: the 10th European congress of stereology and image analysis. Societa Editrice Esculapio-Progetto Leonardo, Italy pp 335–341

van der Sman RGM (2007) Moisture transport during cooking of meat: an analysis based on Flory-Rehner theory. Meat Sci 76(4):730–738

Varzakas T, Escudero I, Economou IG (1999) Estimation of endoglucanase and lysozyme effective diffusion coefficients in polysulphone membranes. J Biotech 72:77–83

Viollaz P, Rovedo CO (1999) Equilibrium sorption isotherms and thermodynamic properties of starch and gluten. J Food Eng 40:287–292

Voogt JA, Hirte A, Meinders MB (2011) Predictive model to describe water migration in cellular solid foods during Storage. J Sci Food Agric 91(14):2537–2543

Vos PT, Labuza TP (1974) Technique for measurement of water activity in the high aw range. J Agric Food Chem 22(2):326–327

Waananen KM, Litchfield JB et al (1993) Classification of drying models for porous solids. Dry Technol 11(1):1–40

Wang CY, Singh RP (1978) Use of variable equilibrium moisture content in modelling rice drying. ASAE Paper 78-6505, ASAE, St. Joseph, MI 49085

Wang Y, Taub IA, Barrett A (2000) Monitoring moisture mobility and migration in model bilayered food systems by NMR, MRI, and mechanical assessment. Proc Scanning 22:67–68

Wang YQ, Wu YP, Zhang HF, Zhang LQ, Wang B, Wang ZF (2004) Free volume of montmorillonite/styrene-butadiene rubber nanocomposites estimated by positron annihilation lifetime spectroscopy. Macromol Rapid Com 25(23):1973–1978

Warin F, Gekas V, Voirin A, Dejmek P (1997) Sugar diffusivity in agar gel/milk bilayer systems. J Food Sci 62:454–456

Watson EL, Bhargava VK (1974) Thin layer studies on wheat. Can Agric Eng 16:18–22

Weglarz WP, Hemelaar MK, Van Der Linder N, Franciosi G, van Dalen C, Windt H, Blonk J, van Duynhoven H, Van As (2008) Real-time mapping of moisture migration in cereal based food systems with Aw contrast by means of MRI. Food Chem 106(4):366–1374

Wesselingh JA, Krishna R (1990) Mass transfer. E. Horwood, New York, 244 pp. ISBN: 0-13-553165-9

Wesselingh JA, Krishna R (2006) Mass transfer in multicomponent mixtures, VSSD. Delft University Press, 329 p. ISBN: 978-90-71301-58-2

Whitaker S (1980) Heat and mass transfer in granular porous media. In: Mujumdar AS (ed) Advances in drying, vol 1. Hemisphere publishing, New York

Whitaker TB, Barre HJ, Hamdy MY (1969) Theoretical and experimental studies of diffusion in spherical bodies with a variable diffusion coefficient. T ASAE 12(5):187–208

Wu Y, Weller CL, Hamouz F, Cuppett SL, Schnepf M (2002) Development and application of multicomponent edible coatings and films: a review. Adv Food Nutr Res 44:347–394

Ziegler GR, MacMillan B, Balcom BJ (2003) Moisture migration in starch molding operations as observed by magnetic resonance imaging. Food Res Int 36:331–340

Zimm BH, Lundberg JL (1956) Sorption of vapors by high polymers. J Phys Chem 60:425–428

Zogzas NP, Maroulis ZB, Marinos Kouris D (1994) Moisture diffusivity methods of experimental determination: a review. Dry Technol 12:483–515

Zogzas NP, Marousis ZB, Marinos Kouris D (1996) Moisture diffusivity data compilation in foodstuffs. Dry Technol 14(10):2225–2253

Index

V. Guillard et al., *Food Structure and Moisture Transfer*, SpringerBriefs in Food,
Health, and Nutrition, DOI: 10.1007/978-1-4614-6342-9, © The Author(s) 2013